A MANAGER'S GUIDE TO ARTIFICIAL INTELLIGENCE CONCEPTS

USEFUL BOOKS BY HARRY KATZAN JR.

Introduction to Computer Science

Computer Systems Organization and Programming

Introduction to Computers and Data Processing

Managing Uncertainty: A Pragmatic Approach

Privacy, Identity and Cloud Computing

Lessons in Artificial Intelligence

Artificial Intelligence

A MANAGER'S GUIDE TO ARTIFICIAL INTELLIGENCE CONCEPTS

Harry Katzan Jr.

AUTHOR'S TRANQUILITY PRESS
ATLANTA, GEORGIA

Copyright © 2023 by Harry Katzan Jr.

All rights reserved. No part of this publication may be reproduced, distributed or transmitted in any form or by any means, including photocopying, recording, or other electronic or mechanical methods, without the prior written permission of the publisher, except in the case of brief quotations embodied in critical reviews and certain other noncommercial uses permitted by copyright law. For permission requests, write to the publisher, addressed "Attention: Permissions Coordinator," at the address below.

Harry Katzan Jr./Author's Tranquility Press
3800 Camp Creek Pkwy. SW Bldg. 1400-116 #1255
Atlanta, GA 30331
www.authorstranquilitypress.com

Publisher's Note: This is not a work of fiction.

Ordering Information:
Quantity sales. Special discounts are available on quantity purchases by corporations, associations, and others. For details, contact the "Special Sales Department" at the address above.

A Manager's Guide to
Artificial Intelligence Concepts/Harry Katzan Jr.
Paperback: 978-1-962492-60-7
eBook: 978-1-962492-61-4

For Margaret, with all my love
now and forever

Introduction

This is an introductory book about Artificial Intelligence intended for managers. It is not for AI specialists. It is too simple for them. As unusual as it sounds, there is a very good reason for its existence. There are existing books on the subject that are very good but are very difficult to read. It's that simple. The concepts are complicated and some require complex math. A typical example is the subject of neural networks.

A person should know something artificial intelligence, because we will be involved if only at the listening level or going into a meeting. Artificial Intelligence is here to stay this time. It's a third try to bring the subject to the forefront. It is here for the future, and it is here to stay because the world needs it. We are in the midst of war, preparation for defense, a dismal economic outlook, crime, killing, and so forth. We need it because if we continue the way we are going, we won't be going anywhere for very long. Practically everything is disarray. Just name them: violence, shady politics, global warming, discrimination, abortion, women's birth problems, men's cancer. What is the solution? It is twofold. Equitably use the information we have and permit the human race to communicate and act in a reasonable manner. Through Artificial Intelligence, often referred to as AI, we have the key to managing our lives on a daily basis – worldwide.

There is a lot going on in the area of AI, as it's called. Young people all over the place are into it. New software is being developed. There is a generative AI system named ChatGPT developed by a company named *OpenAI* that has done a good job with new developments. What is a manager to do.

Our company named IBM is in the field with a system called IBM Watson, that operates out of the cloud is also in the game. In fact, this book can give you some insight into what is going on. It contains three snippets on ChatGPT, IBM Watson, and Data Science that are included at the end of the book.

Instead of business and government leaders having to tell us what they are going to do, we can and will be addressing that subject through AI.

AI is a systems concept, not a single piece of software you can buy at a local store. Can you imagine a kid telling you they have computers and AI, meaning software apps. You will be able to obtain AI capability but it needs to be blended into the infrastructure to its host organization. A good example is the Watson System developed by IBM. It can, will, and has done wonders with design and analytics, but you probably will need some help in employing the computational infrastructure. There are already competition systems but Watson is mentioned because it beat some Jeopardy! champions at their own game.

So, here is what we are providing to you with in this book. A straightforward description of the subject of AI that is easy to read. No more and no less. You can jump into this book at any point and get something out of it.

The book is not expensive and something you would enjoy giving to a relative or friend. The book adheres to the author's principle of no sex, no violence, and no bad language. It is accessible to all readers. This need not be said for a book of this type, but I like saying it.

Enjoy the book if you can, You will be glad, if you only read parts of the book.

The Author

PS: The book contains an AI index. Now you can really enjoy it, because you will have a source of knowledge.

Contents

Introduction .. vii

Part I
Introduction to Artificial Intelligence

Chapter 1 The Basis of Artificial Intelligence 3
 The AI Approach .. 3
 The Scope of AI ... 4
 The Modern Era of Artificial Intelligence 5
 Early Work on the Concept of Artificial Intelligence 6
 Intelligence and Intelligent Systems 7
 Cognitive Technology .. 8
 Virtual Systems and Imagination 9
 Systems Concepts and AI ... 11
Chapter 2 Natural Systems ... 13
 Natural Language Processing 13
 Applications .. 14
 Document Understanding 15
 Semantic Processing .. 19
 Vision ... 20
Chapter 3 Connectionism and Brain Science 23
 Looking at the Brain .. 23
 The Connection Machine 24
Chapter 4 Artificial Intelligence Tools and Technologies 27
 Dimensions of Intelligence 27
 Knowledge Representation 29

Engineering Approach .. 30
Production Systems ... 32
Natural Language, Relations, and Predicate Logic 33
Method ... 34
Blackboard, Reasoning, Inference 35
Forward Chaining ... 36
Backward Chaining .. 37
Explanation Capability .. 38
Chapter 5 Artificial Intelligence Applications 41
AI Technologies ... 41
Natural Language .. 42
Expert Systems .. 43
Expert Systems Architecture,
AI Programming Languages ... 44
Chapter 6 Artificial Intelligence Topics 47
Neural Networks ... 47
Robots and Original Standalone AI Applications,
Technological Singularity .. 48
Machine Intelligence and Learning 50
The Advisor Paradigm .. 51

Part II
Modern Artificial Intelligence

Chapter 7 The Starting Point .. 59
Overview .. 59
The Scope of AI ... 61
The Modern Era of Artificial Intelligence 62
Early Work on the Concept of Artificial Intelligence ... 63
Intelligence and Intelligent Systems 64
Cognitive Technology ... 65

Virtual Systems and Imagination 66
But, Is It intelligent? 67
Systems Concepts and AI 68
Chapter 8 Thinking 71
Chapter 9 Learning 73
Chapter 10 The Advisor Model 75
Chapter 11 Conversational Programs 81
Chapter 12 How We Think 83
Chapter 13 A Quick Look at ChatGPT 85
Chapter 14 A Very Quick Summary of
Artificial Intelligence 89
 Machine Learning, Cognitive Computing 89
 Types of AI, Watson System 90
Chapter 15 Machine Learning 91
 Artificial Intelligence Specifics 91
 Machine Learning Specifics 92
 Artificial Intelligence Applications 93
Chapter 16 Cognitive Computing 95
Chapter 17 Artificial Intelligence Final Report 99
 Summary 99
 The Scope of AI 101
 The Modern Era of Artificial Intelligence 102
 Early Work on the Concept of Artificial Intelligence . 103
 Intelligence and Intelligent Systems,
 Cognitive Technology 105
 Virtual Systems and Imagination 106
 Is It intelligent? 107
 Systems Concepts and AI 108
Detailed Supplement, Thinking 109
 Cognitive Computing 114

Part III
ChatGPT, IBM Watson and Data Science

Chapter 18 What is ChatGPT .. 119
 Introduction .. 119
 Using ChatGPT .. 120
 ChatGPT Internals .. 121
 Social Aspects of ChatGPT, The Author's Opinion 122
Chapter 19 What is IBM Watson ... 125
 Introduction .. 125
 Operational Considerations ... 126
 How in Watson Works .. 127
 Working with IBM .. 127
Chapter 20 What is Data Science ... 129
 Introduction .. 129
 Benefits .. 130
 Technology ... 130

ABOUT THE AUTHOR .. 133
BOOKS BY HARRY KATZAN, JR. 135

Part I

Introduction to Artificial Intelligence

(Chapters 1 through 6)

Chapter 1

The Basis of Artificial Intelligence

Artificial Intelligence is commonly regarded as the science of making machines do things that would require intelligence if performed by humans. Everyone knows this, but it must be mentioned anyway. In addition to being a rather trite definition, the foregoing remark is additionally not intellectually useful. For example, the task of computing the payroll would require human intelligence if done by hand but is regarded as commonplace data processing application. Of course, payroll is not a trivial application and neither are inventory management, online banking, and numerical computing - to name only a few cases. Well then, what characterizes an AI application? Moreover, is there a well-defined dividing line between AI and non-AI? The questions are largely rhetorical since it really doesn't really matter how a particular application is classed. Artificial Intelligence, known as AI, a pragmatic discipline, is characterized more as a way of doing things than a specific well-defined technical concept.

The AI Approach

It is possible to approach Artificial Intelligence from two points of view. Both approaches make use of programs and programming techniques. The first approach is to investigate the general principles of intelligence. The second is to study human thought, in particular.

Those persons engaged in the investigation of the principles of intelligence are normally charged with the development of systems that appear to be intelligent. This activity is commonly regarded as "artificial intelligence," which incorporates both engineering and computer science components.

Those persons engaged in the study of human thought attempt to emulate human mental processes to a lesser or greater degree. This activity can be regarded as a form of "computer simulation," such that the elements of a relevant psychological theory are represented in a computer program. The objective of this approach is to generate psychological theories of human thought. The discipline is generally known as "Cognitive Science."

In reality, the differences between artificial intelligence and cognitive science tend to vary between "not so much" and "quite a lot" - depending upon the complexity of the underlying task. Most applications, as a matter of fact, contain elements from both approaches.

The Scope of AI

It is possible to zoom in on the scope of AI by focusing on the processes involved. At one extreme, the concentration is on the practicalities of doing AI programming, with an emphasis on symbolic programming languages and AI machines. In this context, AI can be regarded as a new way of doing programming. It necessarily follows that hardware/software systems with AI components have the potential for enhanced end-user effectiveness.

At the other extreme, AI could be regarded as the study of intelligent computation. This is a more grandiose and encompassing focus with the objective of building a systematic and encompassing focus with the objective of building a systematic theory of intellectual processes - regardless if they model human thought or not.

It would appear, therefore, that AI is more concerned with intelligence in general and less involved with human thought in particular. Thus, it may be contended that humans and computers are simply two options in the genus of information processing systems.

The Modern Era of Artificial Intelligence

The modern era of artificial intelligence effectively began with the summer conference at Dartmouth College in Hanover, New Hampshire in 1956. The key participants were Shannon from Bell Labs, Minsky from Harvard (later M.I.T.), McCarthy from Dartmouth (later M.I.T. and Stanford), and Simon from Carnegie Tech (renamed Carnegie Mellon). The key results from the conference were twofold:

- It legitimized the notion of AI and brought together a raft of piecemeal research activities.
- The name "Artificial Intelligence" was coined and the name more than anything else has had a profound influence on the future direction of artificial intelligence.

The stars of the conference were Simon, and his associate Allen Newell, who demonstrated the Logic Theorist - the first well-known reasoning program. They preferred the name, "Complex Information Processing," for the new fledging science of the artificial. In the end, Shannon and McCarthy won out with the zippy and provocative name, "artificial intelligence." In all probability, the resulting controversy surrounding the name artificial intelligence served to sustain a certain critical mass of academic interest in the subject - even during periods of sporadic activity and questionable results.

One of the disadvantages of the pioneering AI conference was the simple fact that an elite group of scientists was created that would effectively decide "what AI is and what AI isn't," and how to best achieve it. The end result was that AI became closely aligned with psychology and not with neurophysiology and to a lesser degree with electrical engineering. AI became a software science with the main objective of producing intelligent artifacts. In short, it became a closed group, and this effectively constrained the field to a large degree.

In recent years, the direction of AI research has been altered somewhat by an apparent relationship with brain research and cognitive technology, which is known as the design of joint human-machine cognitive systems. Two obvious fallouts of the new direction are the well-known "Connection Machine," and the computer vision projects at the National Bureau of Standards in the United States. That information is somewhat out of date, but the history gives some insight into what AI is today and where it will be heading.

Early Work on the Concept of Artificial Intelligence

The history of AI essentially goes back to the philosophy of Plato, who wrote that. "All knowledge must be able to be stated in explicit definitions which anyone could apply," thereby eliminating appeals to judgment and intuition. Plato's student Aristotle continued in this noble tradition in the development of the categorical syllogism, which plays an important part in modern logic.

The mathematician Leibnitz attempted to quantify all knowledge and reasoning through an exact algebraic system by which all objects are assigned a unique characteristic number. Using these characteristic numbers, therefore, rules for the combination of problems would be establishes and controversies could be resolved by calculation.

The underlying philosophical idea was conceptually simple: Reduce the whole of human knowledge into a single formal system. The notion of formal representation has become the basis of AI and cognitive science theories since it involves the reduction of the totality of human experience to a set of basic elements that can be glued together in various ways.

To sum up, the philosophical phenomenologists argue that it impossible to subject pure phenomena - i.e., mental acts which give meaning to the world - to formal analysis. Of course, AI people do not agree. They contend that "there is no ghost in the machine," and this is meant to imply that intelligence is a set of well-defined physical processes.

The discussion is reminiscent of the mind/brain controversy and it appears that the AI perspective is that "the mind is what the brain does." Of course, the phenomenologists would reply that the definition of mind exists beyond the physical neurons; it also incorporates the intangible concepts of what the neurons do.

Accordingly, strong AI is defined in the literature as the case wherein an appropriately programmed computer actually is a "mind." Weak AI, on the other hand is the emulation of human intelligence, as we know it.

Intelligence and Intelligent Systems

There seems to be some value in the ongoing debate over the intelligence of AI artifacts. The term "artificial" in artificial intelligence helps us out. One could therefore contend that intelligence is natural if it is biological and artificial otherwise. This conclusion skirts the controversy and frees intellectual energy for more purposeful activity.

The abstract notion of intelligence, therefore, is conceptualized, and natural and artificial intelligence serve as

specific instances. The subjects of understanding and learning could be treated in a similar manner. The productive tasks of identifying the salient aspects of intelligence, understanding, and learning emerge as the combined goal of AI and cognitive science. For example, the concepts of representation and reasoning, to name only two of many, have been studied productively from both artificial and biological viewpoints. Software products that are currently available can be evaluated in the basis of how well they can support the basic AI technologies,

The key question then becomes: How well do natural and artificial systems, as discussed above, match up to the abstract notion of intelligence

Cognitive Technology

Cognitive technology is the set of concepts and techniques for developing joint human-machine cognitive systems. People are obviously interested in cognitive systems because they are goal directed and employ self-knowledge of the environment to monitor, plan, and modify their actions in the pursuit of their goals. In a logical sense, joint human-machine systems can also be classed as being cognitive because of the availability of computational techniques for automating decisions and exercising operational control over physical processes and organizational activities.

Recent advances in heuristic techniques coupled with methods of knowledge representation and automated reasoning have made it possible to couple human cognitive systems with artificial cognitive systems. Accordingly, joint systems in this case would necessarily have the following attributes:

- Be problem driven, rather than technology driven.
- Effective models of underlying processes are needed.

- Control of decision-making processes must be shared between human and artificial components.

Clearly, cognitive technology represents a possible (if not probable) paradigm shift whereby the human self-view can and wake in the not-too-distant future.

Virtual Systems and Imagination

Methods for reasoning in expert and cognitive systems are well defined. Rules and representation effectively solve the problem. There appears to be a set of problems, however, that seem to evade such a simple solution as rules and representation.

A sophisticated model of a cognitive system must incorporate the capability of reasoning about itself or another cognitive system and about the computational facilities that provide the cognition. When a person, for example, is asked to reason about the feelings of the probable response of another person, a set of rules is normally invoked to provide the desired response. If no rule set exists, then a virtual process is engaged that proceeds somewhat as follows:

- The object process is imagined, i.e., you effectively put yourself in the other person's shoes, so to speak.
- The neural inputs are "faked" and the brain responds in somewhat the same manner as it would in real life.
- The result is observed exactly as though it had taken place.

Thus, a sort of simulation of a self-model is employed. This type of analysis might be invoked if someone were asked, for example, how they would feel if they had just received the news they had contracted an incurable disease.

The process, described above, is essentially what an operating system does while controlling the execution of a "guest" operating system. Inputs and outputs are interpreted, but machine code is actually executed.

It necessarily follows that executable models, as suggested here, are as much a form of knowledge as are rules and facts. But, Is It intelligent?

Exploring the issue even further, a sharp borderline between intelligent and non-intelligent behavior, in the abstract sense, probably does not exist. Nevertheless, some essential qualities might be the following:

- To respond to situations very flexibly.
- To take advantage of fortuitous circumstances.
- To make sense out of ambiguous or contradictory messages.
- To recognize the relative importance of different elements of a situation.
- To find similarities between situations despite differences which may separate them.
- To draw distinctions between situations despite similarities which may link them.
- To synthesize new concepts by taking old concepts and putting them together in new ways.
- To come up with ideas that the are novel.

Viewed in this manner, intelligence is a form of computation. Effective intelligence then is a process (perhaps a computer program) and an appropriate machine in which to execute the process.

Systems Concepts and AI

An interesting viewpoint the concerns the evolution of data processing has emerged from the AI business. The task of designing a rule base and an associate fact base is somewhat analogous to designing an information system. Moreover, both kinds of systems appear to evolve in a similar manner. For this analysis, it should be assumed that statements and computational processes (i.e., modules) are the same (or synonymous).

A sensory stimulation is associated with a statement. (Incidentally, this concept is known as associations, wherein a sensation is associated with an idea, and that idea leads to another idea, and so forth. This theory originated with Aristotle and was pursued by Hobbs, Locke, and Mill.). The associations reverberate through the system of statements, whereby a result is finally achieved. The output can viewed as a prediction. Moreover, the system operates according to some form internal laws - such as the laws of mathematics.

When a prediction fails, continuing with the statement analogy, we question the validity of the set of statements. Revisions are normally in order. Since a direct correlation is usually possible between the stimuli and peripheral statements, these statements are preserved from revision. Other statements must bear the brunt of change. The other statements, however, can be regarded as the "frozen middle," since they result from internal laws. The end result that a priority judgment is necessary to change the peripheral statements and change the frozen middle. The priorities of course are in conflict and the preference commonly goes to the revision that disturbs the system the least.

Effectively, incremental changes are made to the system until a total revision is necessary - i.e., a rewrite of the internal including the Laws, the frozen middle, and the peripheral

statements. As a total concept, major revisions serve to simplify a system. It necessarily follows that some attention should be given to systems evolution as a predictive technique.

END OF CHAPTER ONE

Chapter 2

Natural Systems

Developments in artificial intelligence have through the years tended to mirror natural systems. The two main targets in this area are natural language and vision. The field of connectionism is also introduced briefly. Locomotion and robotics are not included in the book.

Natural Language Processing

The use of language in an efficient and effective manner is considered to be an important aspect of human intelligence, and it is therefore no surprise that AI researchers have devoted a considerable amount of attention to the subject. Natural language is a complex area that had its roots in computer linguistics.

The subject of computational linguistics covers the computational mechanisms that form the basis of the use of natural language. The science is a discipline in is own right but lies on the periphery of artificial intelligence, philosophy, classical linguists and psychology. The goals of the field are:

- The investigation of human communications.
- The development of artificial systems with communications capability.

The goals of the field tend to work together. The scientific goal of understanding human language and its use allows us to better use the language and to facilitate the engineering goal to develop systems that process that language. The notion of science and engineering progressing in lock step is not unknown in modern society.

Natural language has evolved as a mechanism for communicating between human minds with all if their functional advantages and processing limitations. Accordingly, there is no underlying reason for computer systems that deal with natural language, or significant subsets of it, to process the language in the same manner that humans do. This is to be expected. Airplanes do not fly like birds, and clothes washers do not wash like people do.

Applications

The main areas of natural language processing are machine translation, document understanding, document generation, and natural language interaction. A fifth area that concerns word processing, word processing, desktop publishing, and electronic mail is also recognized but considered to be adjunct to this discussion.

Machine translation is concerned with the development of computational devices that read a document in one language and produce a document in another language. The ultimate goal is to perform the process in real time – as in a telephone conversation between Japanese and English-speaking persons, each speaking in their language. The science of machine translation is well developed and is applicable to specific domains of discourse – such as an owner's manual for an automobile.

The field has evolved through tree stages. The first step involved single word-by-word translation with a considerable amount of manual pre- and post- editing. The second stage recognized syntactic structure and progressed through a document on a sentence basis. In some cases, an intermediate language was used. In the third and current stage, preference semantics, contextual dependency, and other forms of semantic analyses are used to create a domain of discourse. Current results are very good for certain applications, such as technical reports and reference manuals. Machine translation probably will never do a good job on literature, because of the semantic structure normally associated with varying levels of meaning. A machine translation system can be used as a service facility in a comprehensive management system, and the domain of discourse will expand as technical writers learn to "write for translation."

In the field of professional translation, computer assisted translation (CAT) is currently the norm. Professional organizations, such as Berlitz, use a CAT system to build up a domain-specific vocabulary and can then expect excellent "automatic" translation results on associated projects. A typical application would be the machine translation of legal briefs related to a technical discipline, and a typical domain-specific vocabulary would be 17,000 words. One might conclude, therefore, that machine translation is and has been a successful endeavor, provided that a reasonable definition of the term "success" is kept in mind.

Document Understanding

Document understanding is the process of abstracting the salient aspects of a document, which could range from a telex message to a business report. Clearly, similar techniques machine translation could be used. More specifically, typical applications of an "understanding machine" would be:

- Read and assimilate a document in order to be able to answer questions about it.
- Produce an abstract of a document for library-like purposes.
- Determine the relevant aspects if a document for directing it for further processing.
- Ascertain the salient aspects of a document to alert people that might be interested in it.

In some cases, such as in the processing of a telex message, a sophisticated analysis might not be necessary. For example, a simple content analysis, as in military intelligence or political science, coupled with a relevant profile for redirection is all that is necessary. On the other hand, the "deep structure" of a document could be needed for a question-answering system, where a full-blown semantic analysis would ordinarily be required.

Document generation refers to the process of translating information stored in a formal language into ordinary natural language. Scientific, engineering, and technical artifacts are commonly stored in an internal form to facilitate processing, storage, and retrieval. Engineering drawings and repair procedures are two typical examples. The implication in the above discussion is that text is generated in the traditional form – i.e., printed or displayed on a video display unit. Generation of the spoken word is another option that has potential of the future. Clearly, this capability will be combined with reasoning systems to supply the user with detailed knowledge of how to resolve a particular situation as in the classical case of auto repair.

The primary use of natural language processing so far has been as an end-use interface component is a larger system, such as a database query system. The basic procedure operates somewhat as follows:

- The end user enters a query in natural language form.
- The request is scanned, parsed, and analyzed using natural language processing techniques.
- Ambiguities are resolved through interaction with the user.
- The natural language query facility constructs a query procedure.
- The query procedure is executed by the system.
- The results are returned to the user.

An analogous scheme can be used to control complex equipment and to supply expert advice (such as medical) to the front-end expert system

The modus operandi of natural language interface facilities are threefold:

- To generate executable query procedures, as covered above, that may be referred to as a "loosely coupled" interface to be used with database systems.
- To operate interpretively and generate queries internally from the interface system, and called a "closely coupled" interface to be used with control applications.
- To provide a "menu drive" language, whereby an end-user is supplied an option menu for generating the next word in a query.

The natural language processing process involves the use if a grammar and a lexicon together with a processor that associates input strings with an underlying structure. The grammar, which describes the allowable sentences in the language, and the lexicon, which is an inventory of the words in the universe of discourse and their inflectional, syntactic, and semantic properties, are the inputs to the processor along with the set of input strings. The processor operates through four components:

a parser, a recognizer, an analyzer, and a transducer. The parser operates on a phrase using the grammar and the lexicon. The recognizer accepts the string or rejects it, depending upon the criteria established for denoting well-formed strings. The analyzer associates the input string with its structure, according to the grammar. The transducer associates the string with alternate structures that reflect the domain of discourse. The components are summarized in two sections: syntactic parsing and semantic processing.

Without going into theoretical analysis of the subject matter, it is possible to describe the two most productive methods of syntactic parsing: Augmented Transition Networks (ATNs) and Chart Parsing.

There is a significant amount of expectation inherent in natural language. When a person hears or reads the article "the," for example, an adjective or noun is expected next in line. An ATN is a network of nodes (called states) that reflect the situation. One node denotes the start and one or more nodes serve as a finish. When a word of a given category is encountered, the word is consumed and the system goes into a new state. Overall, an ATN reflects a sentence and sub networks reflect constituents of a sentence, such as noun phrases, verb phrases, and so forth. The process is top-down and internal registers are used to store contextual information and direct the flow. Highly complex grammars can be parsed with ATNs.

Chart parsing is a bottom-up alternative to the use of ATNs; both are based theoretically on a linguistic concept known as Phrase Structure Grammars (PSGs). With chart processing, the text is recognized and assembled word-by-word. Words are combined into syntactic entities, eventually resulting in complete sentences. Varying combinations of words are constructed with the ultimate objective of synthesizing a feasible interpretation.

Chart processing can provide a partial analysis even if a complete grammatical sentence in not available.

Phrases are constructed with chart processing that span increasingly larger sections of a sentence. Accordingly, it may generate alternate interpretations of a sentence without providing any means of deciding the preferred result.

Semantic Processing

The best-known methods of assessing the semantic properties of a sentence are case grammars and contextual analysis. The objective is obvious: to control the level of ambiguity in natural language.

A case grammar can be viewed as either a detailed syntax of a generalized semantics based on the verb of a phrase or sentence. The value of case grammars is that for most verbs, a frame can be defined. The slots of the frame reflect the subject of the verb, object, and so forth. Case grammars are sometimes known as "case frames."

Conceptual analysis (CA) takes the notion of case grammar one step further by defining 11 distinct primitive actions, normally in discourse. Two common examples of CA primitives are ATRANS and PTRANS. ATRANS represents an abstract relation, such as the transfer of ownership, and PTRANS is used to denote the physical transfer of an object. Thus two equivalent sentences, such as:

John sold Mary a book
Mary bought a book from John

Would be represented is the same way using conceptual analysis using the primitive ATRANS. Similarly, the sentence "John went home" would be described with PTRANS.

In conceptual analysis, each primitive has "expectations" that are expected, so in fact, they can play a key role in a larger unit of description called a script. Most everyday actions, such as going to a restaurant or the post office can be modeled with an appropriate script.

There is a high level of activity in the DP world and in AI circles on the topic of natural language processing. The reasons are clear. Natural language is understood by practically everybody, and it is self-updating. The bandwidth of natural language is high, and it presents a low initial barrier to end users of DP systems. Natural language systems lend to human adaption as writers and end users quickly learn the sentences that succeed, in a figurative sense.

Vision

Recent programs in locomotion and robotics have created a new need for a close look at vision systems. After years of only marginal success, AI researchers have agreed that the problem is much more difficult than they originally thought. Existing industrial vision systems look at high contrast scenes and compare the resulting lines and dots to existed images. These systems use a hierarchy of representations that allow processing of visual information in many intermediate stages to traverse the conceptual distance between input signals and cognitive symbols.

After a quarter of a century or more of development, vision research has become a major subfield of AI. Early systems were concerned with image recognition in the forms of edge elements and primitive data objects, and image manipulation such as scaling and rotation. A cognitive approach subsequently surfaced that emphasized symbolic processing and knowledge

representation. The later direction led to notion of abstraction and a vision hierarchy.

AI components in vision systems utilize texture and motion to key into semantic data structures that promote a "context of vision," analogous to the methods of conceptual analysis. The research is close to psychology, as was the case with natural language processing and a "model driven" approach that is currently fashionable.

Using the human brain as a model, scientists outside of the mainstream of traditional AI have built vision systems that combine technologies from electronics, robotics, and computer science. The key to success would appear to be the development of special hardware units for vision.

END OF CHAPTER TWO

Chapter 3

Connectionism and Brain Science

One of the most severe limitations of natural language and vision systems is computation speed. Von Neumann machines are too slow for sophisticated practical work, and most AI researchers agree that parallelism is the answer. There are several well-known ways of achieving parallelism; the emphasis here is on a rebirth of area of AI known as connectionism. Parallelism is surveyed briefly as a related technology.

Looking at the Brain

The human brain has traditionally been regarded as a sophisticated switching network consisting of billions of processing elements, known as neurons that communicate through electrical impulses. Since the 1970s, however, neuroscientists have determined that neurons work together in large groups and interact through electromagnetic fields. This sheds a new light on cooperative behavior, since the synchronous behavior of thousands of neurons that are continually and instantaneously reflected back to affect subsequent activity. This is clearly a new direction in associative behavior and has led computer designers to pursue a new type of computer – the Connection Machine. Even though, the notion of a connection machine has not been totally accepted by the business and computer communities, the concept is certainly worthy of note.

There is some history and value to this direction of research. Scientists at Carnegie-Mellon University have developed a software simulation of a novel computer architecture known as the Boltzmann Machine. Each processor is connected to hundreds of other and is either "on" or "off" at any moment. The links between a processor and its neighbors area assigned numerical "weights" that can be either positive or negative. To decide whether to change its state, each processor sums the weights of the inputs it receives from other currently active processors and compares this sum with a threshold value. It then calculates its own state accordingly to a probabilistic decision rule. If the sum is above the threshold value, the processor the processor turns on half of the time. Because the activity of the individual processors is probabilistic, the behavior of the network as a whole can be analyzed with the same statistical formulas that describe the kinetic behavior of gases. For this reason, the simulated machine was named after Ludwig Boltzmann (1844-1906), as Austrian physicist whose research on gas thermodynamics led to the field of statistical mechanics. Information in the Boltzmann Machine is stored not in the individual processors but in their connections, which are represented mathematically by the numerical weights. Machines of this gender are relatively easy to conceptualize and serve was term projects for students. Programs are relatively short to write and give the students a keen sense of accomplishment in the field of artificial intelligence. Petri Nets are a commonly used production vehicle for this type of project.

The Connection Machine

The basic computing speed of the brain is a few milliseconds, so that a reasonably complex task, such as recognizing a face, which takes about half of a second, is computed in roughly 100 steps.

The number of sequential time steps to perform the same task in an abstract AI system runs into the millions. Clearly, the brain is a highly parallel device. Moreover, large symbolic structures cannot be passed through the neural structure in binary form with the 100-step restriction. Based on analyses, such as these, a "connection" model of the brain is hypothesized that employs connections among the units that go far beyond the neural structure of the brain in its traditional form.

Parallelism can be designed into a computer system at various levels. Low-level parallelism is achieved by pipelining techniques, interleaved storage, and parallel functional units. All of these techniques hare transparent to conventional in conventional software. A higher level of parallelism, on the other hand, normally involves the interaction of several closely coupled processors connected in the form of an array or a network. Thus, the multiple processors can cooperate on a computational task but operate in tandem.

Two methodologies stand out in the above discussion: dataflow computers and connection machines. A salient characteristic of traditional and parallel computers, as discussed thus far, is that they are process driven. What this means is that procedures, - i.e., machine code, normally written as statements in some programming language – essentially run the show. Moreover, processes request data when they need it and blindly go about doing what they do. The problem is obvious. Data is not always readily available when needed, and this causes computational delays and thwarts effective parallelism.

In a dataflow computer, the flow of data is optimized. Each processor receives data packets from other processors. When a process has the data it needs, it performs a calculation and sends the results on to another processor. Thus, each processor is supplied with a storage area with instructions on what do with

the data and associated interconnection ports. Data flow programs are written as a "map" describing the flow of computation. Software is used to convert a high-level description into instructions for the constituent processors.

The Connection Machine no longer exists but is a useful computer but its design supplements the story of artificial intelligence. It is a non-von Neumann computer, often called non-Von, wherein each processor is connected to the remainder through a 4,096-switching station and 24,576 trunk lines. The Connection Machine has scanned three months of Reuters News – 16,00 articles in all-in 1/20 of a second. The operational speed of the computer is in excess of one billion instructions per second equivalent to a Cray supercomputer at a fraction the cost.

END OF CHAPTER THREE

Chapter 4

Artificial Intelligence Tools and Technologies

In spite of a considerable amount of discussion, the notion of intelligence appears to be one of those things that is difficult to define bit is readily recognizable in practice. When does an object possess intelligence? Clearly, there are fast minds, to use the term loosely, and slow minds. Moreover, there is more than one type of mind. There are differences between adults, between children, and between and adults and children. There are differences between humans, dogs, monkeys, and other animals. There are cross-cultural differences, and there are differences between intelligent machines. There doesn't seem to be a clear boundary between what is intelligent and what is not. It is possible, on the other hand, that the field of Computer Science has provided us with a unique approach to the quest for intelligence. The concept of a virtual machine, defined as an abstract structure that can support diverse computational states and processes, is a useful paradigm for identifying the boundaries of artificial intelligence.

Dimensions of Intelligence

A virtual machine can reside in a physical machine without having to be that machine. Similarly, the same virtual machine can be installed on two different physical machines, and two different virtual machines can reside in the same physical machine. It follows

that different virtual machines can have different properties.

In the realm of AI, one virtual machine can play chess, another can support oil prospecting, another can perform as a medical assistant, and yet another function as a financial advisor. Then if we look at the space of all such virtual machines, we get a realistic picture of what it is that artificial intelligence is supposed to be. There is no clear boundary between possessing intelligence and not possessing intelligence. The dimensions of intelligence do not form a smooth continuum but are marked with many discontinuities. Some AI virtual machines could appear to exhibit a reasonable level of intelligence, such as an expert system, while others, such a pick-and-pace robots, hardly seem deserving of the term "intelligent."

Without making a claim to absolute completeness, it appears as though the following AI technologies could adequately represent the dimensions of intelligence in a hypothetical AI virtual machine: Knowledge representation (KR), reasoning (R), vision (V), pattern recognition (PD), natural language processing (NLP), and speech processing (SP).

As the dimensions multiply, an AI system gives the appearance of being more intelligent, and a quick estimate is that the effect is exponential. The field of artificial intelligence is commonly viewed in terms of its applications and not in its technologies. It is common to read of AI applications with interesting names, such as MYCIN, HACKER, and ACTOR. The list is old but the point has been made. This is unfortunate since the technology of AI is a knowledge domain.

In reality, the computational facilities often used in AI work is means of solving a problem and not a complete set of AI applications. Thus, the domain of AI tools spans the set of applications. The two most significant technologies are knowledge representation and reasoning.

Knowledge Representation

There is a concept in AI known as the primacy of knowledge. What this implies is that the ability of a system to perform a task is a function of the knowledge that the system has about the task – regardless of how cleverly the system reasons about it. In this regard, AI is an empirical art and not a theory. It is the engineering procedure of building tools, evaluating them, tearing them down, and repeating the process. Currently, there are elements of a theory in AI and eventually it will have a theory as in Biology. An emphasis on knowledge representation (KR) makes sense, because most of the world's work is thinking, which is composed of problem solving, concept formulation, and knowledge processing.

In its traditional form, knowledge is more static than data. Two varieties of knowledge are commonly recognized: factual knowledge, which is acquired through learning, and heuristic knowledge, which is gained through experience. It is not a joke that heuristic knowledge is known as "The Art of Good Guessing." It follows that capturing the art of good guessing is commonly regarded as knowledge engineering.

Human knowledge is commonly regarded as passive information recorded in books and periodicals and facts stored in the minds of individuals. Through their existence, many structures created by society are implicit in the design of those structures. In the field of computer technology, computer programs use algorithms that operate on stored data. An expert system, commonly known as a knowledge system, is a program that operates on a class of stored data. Collectively, an expert system is comprised of the information and a set of algorithms.

When an expert system is created, the main task is the modeling of the expert. People have only limited capability to construct long chains of deductions and have only limited

knowledge, at any one point in time. This is a form of limited human rationality.

Because of limited human rationality, representation is a key issue, because it spans the various steps of the inference process. One categorization of knowledge representation is that it is a stylized version of the world. It is stylized because the method(s) employed depend of the objects to be represented and the expected processing that will be performed on the objects that are represented.

Engineering Approach

In the human brain, knowledge is stored as neurochemical structures, electromagnetic fields, electrical impulses, and probably other unknown methods. I will take scientists many years to sort out precisely how the brain stores information and, in fact, how it uses that information. As a practical endeavor, an attempt to model the brain is simply not effective. AI researchers have alternately selected to "engineer" a few methods that work for the kinds of problems encountered in today's world. As a practical example, airplanes do not fly like birds. Engineering is necessary.

The engineering approach to knowledge representation would first involve a look at the problem domain. In the case of artificial intelligence, the basic needs include taxonomies, inheritance, constraints, and uncertainty.

Taxonomy refers to classification, as in the following example:

Cows are mammals.
Mammals are vertebrates.
Mammals suckle their young.
Vertebrates have backbones.

Taxonomy, in its most general sense, is the concept through which knowledge is stored.

Inheritance, albeit related to taxonomy, involves the capability of obtaining attribute values through a hierarchical structure without explicitly stating them, as in the following case from propositional logic:

> All birds have feathers.
> Tweety is a bird.
> Therefore, Tweety has feathers.

In this case, an object inherited an attribute value from a more general class. Inheritance provides more than just a means of picking up attribute values from a superclass. It encourages and facilitates effective organization of knowledge – something the brain does automatically in its magical ways.

The notion of constraint refers to the process by which a value is known if its associated values are known. In the case of Ohm's Law of the form:

$$V=IR$$

for electrical resistors, a third value can readily be computed if the other two values are known. In advanced knowledge systems, a constraint value is encoded in the slot of a frame and is available through a method attached to that slot.

The concept of uncertainty is fundamental to all knowledge. Heretofore in AI and expert systems, a tacit assumption has been made that all knowledge statements are true. For special cases, such as medical diagnosis, certainty factors or belief functions were simply 'tacked on.' A modern look at knowledge representation should be based on the uncertainty that is inherent in most knowledge systems.

Production Systems

The most widely and extensively tested means of representing an expert system is to use the concept of a production system developed in the U.S.A. in the 1970s and applied in one form or another in most large-scale expert systems. A production system is used for the representation of knowledge and for AI reasoning.

A production system is a set of condition-action pairs, called production rules that are used to represent an inference procedure. Accordingly, the form of a production rule is:

IF <condition> THEN <action>

The "conditions" are ascertained from input values that are facts in the knowledge base and from previous "actions."

Associated with an inference procedure is a short-term memory, termed the context, the long-term memory called the knowledge base, and a set of rules and a rule interpreter, sometimes known as the inference engine. In French AI, the inference engine is called the motor, with the obvious analogy to the functioning of an automobile. The motor drives an automobile in the same manner that an inference engine drives an expert system.

The inference engine, which is actually a software module, has access to the rule base, the context, and the knowledge base. The engine operates by finding all rules for which the condition part is true. A control strategy is applied to select the rule to be found, and the process is again repeated. The "action" can take the form of an end-user interaction, the execution of a hardware or software function, or the computation of a value that is added to the context.

A production system can be used to store knowledge in a way that is functionally analogous to tables, structures, hierarchies,

nets, frames, and so forth. Each element of knowledge is represented is represented by a rule. A typical production rule of this type might be:

IF age = 30 THEN risk level is 5

Clearly, this information could have been stored equally well in a table. It is a matter of record that in expert systems with thousands of production rules, many of the rules are used only to store knowledge.

The basic methods of knowledge representation are natural language, predicate logic, frames, semantic nets, methods, and blackboards. Most methods are use in combination. Conceptual dependency graphs are considered to be a special case; they have been introduced previously.

Natural Language, Relations, and Predicate Logic

A natural language representation of knowledge would take the form of a normal sentence, such as

John gave Mary the book

This knowledge could be represented in an unambiguous manner is the form of a relation, such as

GIVE (JOHN, MARY, BOOK)

Similarly, the following items represent the same ideas:

John is a programmer
OCCUPATION (JOHN, PROGRAMMER)

Mary is a lawyer
OCCUPATION (MARY, LAWYER)

and

John's address is 37 Maple Street
ADDRESS (JOHN, 37-MAPLE)

Items in parenthesis are attribute values. Items, such as GIVE, OCCUPATION, and ADDRESS are relations, as in database theory. In predicate logic relations are known as "predicates."

Expert systems and logic programming languages can use predicates in the reasoning process.

A semantic net is a relational model among objects, classes, and attributes, as suggested by the following diagram:

Tweety IS-A Canary IS-A Bird HAS-PART Wings

In this case, "IS-A" and "HAS-PART" are relations, and the other entities are an object (Tweety), classes (Canary and Bird), and an attribute (Wings). Clearly, objects may have attributes, attributes can have attributes, and so forth.

Semantic nets are particularly useful for expert systems, because entities in a net can be described by frames and can inherit attribute values from other entities in the net. Moreover, objects may participate in several relations, providing complete generality as a representative scheme.

Method

A method is a computational procedure attached to a slot, which is a data structure in AI. A reference to a slot causes the associated

procedure to be invoked, which computes the needed value. The notion of a method is a necessary concept in AI. It appears that only two facilities are necessary to allow and expert system to work: a frame and at least one method. A knowledge base is stored in a frame with hierarchies, so that inference can be performed by the methods.

Blackboard

A blackboard is a knowledge structure that facilitates the concurrent, and perhaps cooperative, activity of multiple reasoning agents. The current state of a problem is represented in its blackboard. Each reasoning agent possesses a specialty and applies his or her area of expertise to an appropriate part of the problem as a solution unfolds on the blackboard.

Reasoning

Reasoning is commonly understood to be the process of inferring one fact from another. In reality, inference can come in three forms: deduction, abduction, and induction. Inference methods are inherent in forward and backward chaining.

Inference

The basis of production systems and much of the current crop of expert systems is the process of deduction. Deduction takes the form:

 IF a THEN b
 a is true

 Infer that b is true

This is a rule in logic called Modus Ponens, which produces "legal inference." Abduction takes the form:

IF a THEN b
b is true

Infer that a is true

This is a rule in logic called Modus Tollens. Abduction does not product "legal inference," and is a means of generating a questionable explanation of observation. Abduction can lead to the wrong conclusion but is very useful and often necessary. Abduction is often pertinent to medical diagnosis. The final form of inference is induction. Induction is essentially the search for patterns in observed data and takes the form:

P a is true
P b is true
...

Infer for all x, P x is true

Induction reflects the process of learning, which is in its infancy in AI.

Forward Chaining

One method of using a production or rule-based system is to reason from first principles or premises, as they are normally known. This process, known as forward chaining, starts with a fact and then finds a rule whose premise is satisfied by the fact.

One means of conceptualizing forward chaining is that it is a means of responding to new facts and simulating actions. Accordingly, forward chaining operates somewhat a s follows:

- The interpreter looks at the first of the new facts.
- All of the rules are identified to which that fact is a premise.
- A check is made whether all the premises for the flagged rules are verified.
- For all verified rules, the specified actions are taken, often by updating the context.
- The process is repeated with the next item from the set of facts.

The interpreter in question is the "production rule interpreter," which is commonly a part of an expert system environment.

The notion of a control strategy frequently arises because the order in which the interpreter runs through the production rules can determine the results that are obtained. Some of the strategies commonly mentioned are to apply the last rule used, the highest priority rule, and the rule that refers to the value most recently added to the context.

Forward chaining is particularly useful for AI problems that involve diagnosis systems and "what to do" applications.

Backward Chaining

In contradistinction to forward chaining, backward chaining starts with a conclusion to be established and works toward the premises or conditions that would cause it to happen. Backward chaining is tantamount to breaking a problem down into sub-goals, the sub-goals into lower sub-goals, and so forth.

Backward chaining is particularly useful for AI applications in which a goal of some kind is to be reached and a good strategy is needed to get there. During backward chaining, a hypothesis is established, perhaps it is even selected by the AI system itself. The interpreter scans its rule base to see if its conditions are

correctly established as facts. If so, then the hypothesis is proved. If the conditions of that rule occur as conclusions (i.e., actions) of another rule, then the conditions for that rule are similarly inspected. Thus, the system works backwards until the truth or falsity of all relevant conditions have been ascertained.

The process of backward chaining is summarize as follows:

- A hypothesis to be satisfied is selected.
- A rule for this hypothesis is found.
- A check is made to see if the conditions for this rule are true.
- It any of the conditions is a conclusion to some other rule, then backward chaining, as described in this list, is applied to each of those rules and the process is repeated until the set of initial conditions are reached.

Forward and backward chaining can utilize the same knowledge base, since the inference process is independent of the production rule structure.

Explanation Capability

It is normally the case that an expert system is designed to interact with an end user. Certainly, exceptions exist as with expert systems that are designed and developed to control things, such as computer systems and airplanes.

An expert system can interact with the end user in one or more of the following instances:

- For entering facts into the contextual memory
- For stating hypotheses, as in backward chaining
- In response to a query to the user
- To invoke the explanation capability

Items one through three are straightforward and could normally be expected in an information system of any kind. The explanation capability Is unique.

The explanation capability permits an expert system to run through the chain of reasoning used in an inference process and effectively explain its actions. This is particularly useful for training environments and in critical systems – such as medical diagnosis.

Many expert systems are justified solely on the basis of the explanation facility. The underlying thinking is that if an expect system turns out to be unusable by domain experts, it can always be used for training purposes.

END OF CHAPTER FOUR

Chapter 5

Artificial Intelligence Applications

It is usefully to summarize developments in AI technology prior to covering AI applications. The major areas of developments are:

- New data structures for representing things about the world
- New algorithms to represent facts
- Advanced programming environments
- New computer architectures
- Project management
- Enhanced methods for human/computer interaction
- Special applications of AI technology

The developments do not guarantee success, but the lack of them may be associated with failure. The areas of development are: reliable service, user interface, complete toolbox, and the application must demonstrate intelligence.

AI Technologies

The AI technologies mentioned previously include:

- Robotics,
- Vision
- Pattern recognition
- Speech
- Natural languages
- Complexity interface

The various technologies are used primarily in natural language technologies, and expert systems.

Natural Language

There is not much in the area of natural language that is relevant to the current discussion of artificial intelligence. Robotics and speech are superfluous at this time. Vision and pattern recognition are questionable. The technology exists for well-defined jobs, such as visual processing of securities, but is available only as standalone systems.

In modern AI, natural language is well represented.

Natural language is almost as problematical, but there are three general applications that are relevant:

- A natural language interface to a database system
- Analysis of a document composed in natural language for subsequent analysis or rerouting
- A machine translation package as part of a comprehensive management information system

Question-and-answer systems and natural language interfaces to expert systems have been created in the laboratory but are not viable products at this time.

The overall lack of success by the AI community with natural language processing brings up an important question: Is natural language really natural for the computer? For DP applications, as we currently know them, the answer is clearly 'no.' Moreover, natural language has a narrow bandwidth and should be supplanted by high bandwidth systems based on high resolution graphs and enhanced visual systems.

Expert Systems

The subject of expert systems is commonly regarded as a branch of artificial intelligence based on software technology, even though specially constructed hardware may be used in the process. An expert system is essentially a computer program; in fact, the dependence on programming is so great that artificial intelligence is sometimes viewed as a new way of doing programming.

An expert system uses explicitly represented knowledge and computational inference procedures to solve problems that normally require human knowledge and judgment. The subject matter to which an expert system is applied is referred to as its domain. Thus, knowledge relevant to a given task is known as domain knowledge and the associated human specialist is called a domain expert

In conventional computer programs, the knowledge and control are intermixed. In an expert system, the knowledge is in the form of facts, and rules are distinct from the mechanism that uses the knowledge to draw inferences. Thus, the behavior of an expert system can be changed through modifications of the facts or the rules.

Expert systems are based on a symbolic representation of real-world objects and symbolic inferences and are particularity useful in three cases. First, human expertise is not always available. Second, expertise can be applied more uniformly and more rapidly than with a network of experts. Lastly, an expert system provides a common knowledge base that can evolve dynamically as the system is used for not only one but many applications.

It is customary to leverage the cost of an expert system against the cost of a human export, and in many cases, the

concept of an expert assistant can be employed to more profitably utilize the skills of the human. Another key point is that an expert system can normally explain its behavior and can accordingly be used for justification and training.

Expert Systems Architecture

As covered previously, the application software for an export system must contain a knowledge base, and inference engine, and an end-user interface. In fact, any worthwhile software must contain the three elements. The names of the pieces may be different, but the idea must be the same. This is the most important element in this book. In AI, the generic software entities are normally implemented through the following hardware and software products:

- An AI shell
- An AI language
- An AI environment

The AI shell incorporates the inference engine and implicitly specifies the method of knowledge representation. The AI language is the vehicle through which the AI shell and knowledge base are implemented. The AI environment provides the method of display, bit-mapped graphics, and the editor. One typical AI shell with a historical perspective is KEE for knowledge engineering environment. Its AI language is LISP, and the typical environment is the Lambda LISP Machine.

AI Programming Languages

The two most well-known AI programming languages are LISP and Prolog. LISP refers to list processing, and Prolog refers to programming in logic. Both languages are exceedingly powerful

and are reasonably easy to learn. Python is commonly used in modern AI.

END OF CHAPTER FIVE

Chapter 6

Artificial Intelligence Topics

Exactly what constitutes AI is still an open item. Two types of AI have been identified: Weak AI and Strong AI. Weak AI focuses on specific problems, such as specific games, giving directions, and specific medical diagnosis. Strong AI focuses on how humans solve problems and how to apply that knowledge in a computer environment. With weak AI, a set of programs is told how to do a specific task. With strong AI, the computer determines what it has to do to be learning. Weak AI exists today. Strong AI is under development. As covered in the first section of this book, strong AI involves the development of algorithms that can solve a class of problems usually involving a collection of data sets. Expert systems are a typical example of strong AI. If the development of algorithms involves expert systems in its most general sense, the methods involve modeling of how the brain functions and is known as deep learning, usually implements as neural networks. With neural networks, the software takes input and generates output with an algorithm that runs through a series of networks – such like the human brain functions.

Neural Networks

A typical neural network is a connected set of nodes organized somewhat like the human brain. They are used to recognize relationship s between relatively large sets of data. Typical

applications of neural networks include: medical diagnosis, controlling robots, visual recognition, face recognition, running complex mechanical equipment, controlling complex agriculture, banking and portfolio management, learning and playing board games, racist history, military strategy, dieting and weight control, cleaning, painting, and just about every activity usually performed by human beings.

Robots and Original Standalone AI Applications

The Google Corporation has invested in an extensive program in Artificial Intelligence that employs technology in all areas of the discipline. There is no limit to Google's achievement, that includes neural networks, robot technology, control facilities, art and design methods, and modern biological systems. Here are some of the examples: training and learning, learn how to detect and play electronic games, creating and understanding stories, prediction of human behavior, racist behavior, spacecraft and military behavior, early warning and control systems, cancer detection, medical drug design, Improved MRI scans, genetics and autism, interspace behavior and control, and warming and agriculture. There are, of course, additional areas of investigation.

Technological Singularity

Some artificial intelligence researchers have expressed the idea that as AI systems expanded in scope and flexibility, they would eventually take over the human civilization. AI systems are currently designed and built created by humans, but as the systems become smarter, they would be able to design future AI systems that would be able to build systems that would surpass human intelligence.

Then, those "smart" systems could run away with civilization. They would be beyond the control of their human creators.

Many AI people think that a runaway set of intelligent systems taking over their own design would be totally ridiculous. Large software systems contain millions of lines of code and never get bug-free versions in their lifetimes. Some software design and builders, such as IBM, have exotically developed systems that produce nearly bug free code, but still need the end users to help in the checkout process. Some software develop systems with bugs in them and rely primarily on the end users to find errors.

The point when artificial intelligence systems take over the world is known as the singularity or more precisely the technological singularity. Some authors have even written books on the subject. However, the subject is just an idea, since no one has shown how to create the singularity. Polls of computer scientists have been made to determine just how feasible that idea is. Some well-known that at the singularity would result in human extinction. Four polls in the 2012 and 2013 time period were made and there was a one in two chance that we would have a singularity by 2050.

To start out, one would need a pretty large computer to achieve a singularity and that would be quite an achievement in a society where computers can barely recognize a cow. It's a humorous endeavor to think about a singularity and apply Moore's Law, which states that computer power doubles every ten years. This may not be completely accurate since the source is Wiki. (https://en_m.wikipedia.org/wiki/Technological_singularity.) Actually, the entire article is based on this Wiki report.

Several philosophers, computer people, and writers have state that there is no reason to believe in a artificial intelligence singularity, and they essentially doubt that will happen. (See the Wiki for comments by such academics.)

Machine Intelligence and Learning

Machine intelligence and learning reflects a computer system that exhibits behavior that would be classified as intelligent if performed by a human agent. That computer system would be known as an intelligent agent, and that field of study is termed artificial intelligence. Some researchers refer to AI, as it is usually called, as that which hasn't been done yet, even though that term doesn't seem to make sense in today's computer world. That would imply that an intelligent agent should, in fact, perform a certain amount of learning.

Some researchers feel as though AI can be achieved when human intelligence can be described in sufficient detail that a machine can be made to simulate it. One of the early artificial intelligent applications is "compute chess" wherein the program emulates a human chess player. Some AI applications can learn how to play a game and win by learning how to operate without human intervention.

In general, AI researchers feel that any problem that can be solved by a human can be developed as an AI application, and the challenger is to have the AI program learn by itself how to play. That is why machine learning is one of the most important concepts in artificial intelligence.

Other disciplines can be related to machine intelligence, such as advanced statistical techniques, which is termed deep learning. There are several topics related to machine intelligence and learning:

- Reasoning and problem solving
- Knowledge representation
- Knowledge engineering
- Qualification
- Natural language
- Creativity
- General intelligence

51 | A MANAGER'S GUIDE TO ARTIFICIAL INTELLIGENCE CONCEPTS

The various subjects are held together through a Web Ontology Language that can be used to represent what exists.

AI is relevant to any task that can demonstrate intelligence. As before, when a term (or technique) reaches mainstream, it is no longer considered artificial intelligence. This is the AI effect.

Here are some worthy applications: Healthcare, Automotive, Finance, and Video Games. There are three basic questions:

- Is artificial general intelligence possible? Can a machine solve any problem that a human can solve?
- Are intelligent machines dangerous?
- Can a machine be intelligent? Can it "think?"

The questions are answered in
https://en.m.wikipedia.org/wiki/Artificial_intelligence

The Advisor Paradigm

An expert system is characterized by a database of facts and a logical engine for exploiting the facts in a purposeful manner. The logical engine takes good advantage of the available information so that the output of the system is considerably more than what would ordinarily be implied strictly by the input. The data and the inference programs give the system an air of being an expert.

One of the most illustrative examples of an expert system is the well-known wine advisor. The user enters a brief description of an entrée, such as chicken with cream sauce, and the system yields a small wine selection, such as Chardonnay or Montrachet – depending of course upon the current contents of the wine cellar. Naturally, there is more to the story. The entrée has certain characteristics that are determined by a database search. The

characteristics are then used by an inference mechanism during a "run" of the system to generate the attributes of a suitable wine, such as color, taste, and strength (i.e., light, heavy, etc.) In a second database operation, the wine attributes are mapped into a "variety" of wine. Finally, a selection is made from the database that represents the contents of the wine cellar.

Similar advisors have been constructed in the financial community and in the nuclear industry – to name only a few. In an investment advisor, for example, an investor's profile is mapped into a well-defined, but much larger, set of investment characteristics using a database operation. Again, an inference mechanism is used; in this case, it generates a portfolio structure that serves as input to a database search operation yielding a set of investment options. A typical portfolio structure might be: 60% equities, 35% bonds, and 5% gold. The output set would be a small selection of instruments in selected currencies.

In the nuclear industry, a collection of physical events occurring in an emergency situation would be translated by an expert system into a set of instructions to an operator. As before, the processing sequence includes an initial database search operation, an inference procedure, and then a final database selection to generate the needed instructions. The examples typify the "advisor paradigm."

The advisor paradigm is characterized by the following set of actions occurring in essentially the order given:

- A relatively small set of input values are entered into the system via and end user interaction or as the result of a physical event.
- The input values are expanded into the particulars for a certain problem domain via a database search process.
- The particulars are mapped into a set of descriptors via an inference mechanism that can take the form of a set of if-

then rules and a logical engine, a neural network, a pattern matching operation, a mathematical function, or possibly logic program.
- The descriptors are used as key values in a subsequent database search operation to yield the results for that problem.

Clearly, the salient properties of the advisor paradigm are the initial database operation, the inference (or mapping) process, and the final database search. The two database operations are obviously to different structures. Each of these components can be replicated, as required by a particular problem and its problem solver, but the general structure of the paradigm is evident without necessarily going into ore detail. In fact, most expert systems operate in precisely this manner, even though the emphasis has conventionally been on the rules and knowledge-based parts of the model.

The advisor paradigm is independent of the mode of representation; this fact is tacitly understood and is inherent in the notion of a paradigm. Representation is dependent upon a specific instantiation of the model. If the cells in a spreadsheet application are the source of information, then the retrieval and inference operations are related to that framework. If the inference processes are implemented as LISP or Prolog programs, then the particulars and descriptors are constrained by the corresponding ad hoc data structures. To sum up, the method of representation is determined by the underlying problem domain and computational framework. This explains to a limited degree why a lot more "advisors" are not floating around in the industrial, academic, and governmental communities. The ultimate objective is clear: to make the benefits of the advisor paradigm available in a general-purpose format.

Another consideration that complicates the picture is the

existence of uncertainty in most areas dealings with human affairs. Uncertainty has been the Achilles heel of the artificial intelligence community, since a practical system for problem solving should take the uncertainty of information or the uncertainty of the reasoning process into consideration. Naturally, we would like to have the mechanisms in place for dealing effectively with both forms of uncertainty, even though all of the features may not be needed for a particular problem situation. It is indeed one thing to map an entrée of beef into a wine descriptor in a deterministic sense, such as:

beef -> strong red meat -> dry heavy red wine -> ...

and quite another to introduce the notion of uncertainty, as in:

small lump -> (((healthy, tests))),0.6),((healthy, tests, sick),0.4))
->... -> ((home, office).0.56)

The notation in the latter example is intended to demonstrate that a small lump is mapped into the conditions of healthy or tests with a certainty of 0.6 and into the conditions of healthy or tests or sick with a certainty factor of 0.4. Total comprehension here is not expected, but is given to indicate that uncertainty is complicated. The stage is obviously set. A system for managing uncertainty.

In the wine and medical examples, there is relatively little risk in the advisory process, and gives us the opportunity of focusing on the initial database operation. The process represents a translation between the external work=ld of the end user to the internal world of the automated problem solver In the wine advisor, for example, the indicator of beef is mapped into a set of attributes with which the expert system can work, as in:

beef -> strong red meat

or in its generic form:

> indicator -> meaning

The notion of an indicator is well-established in organized society in the form of symbols, variables, synonyms, and abbreviations. It is commonplace to have an indicator, such as successful, that is associated with a substantial set of characteristics, such as aggressive, confident, intelligent, and focused. Indicators can have different meanings to different people' however, which preserves the role of the expert. In the absence of uncertainly, and indicator has a definite meaning. In the presence of uncertainty, an indicator might mean something, and the degree to which an expert believes the assertion is known as its certainty value, or as its the certainty factor, as it is sometimes called. Thus, the indicator in the presence of uncertainty is a clue to the validity of an assertion; it becomes evidence when combined with its meaning and a certainty value, or in its generic form:

> Evidence: indicator -> meaning with certainly value.

An obvious consideration applies to the notion of evidence. As in common everyday usage, a fact "is entered as evidence" only when it is relevant to the case under consideration. An analogous caveat applies to the methods contained here, and a means is presented for constraining the relevancy of information.

END OF CHAPTER SIX AND PART ONE

Part II

Modern Artificial Intelligence

(Chapters 7 through 17)

Chapter 7

The Starting Point

Overview

The question, "Can a machine think?" is one that has been debated for some time now and the question is no likely to be answered in this book. However, the subject is fruitful when considering "what a computer can do."

There are various opinions on the subject. Some say that thinking is an activity that is peculiar to human beings. Accordingly, machines cannot think. Although thought as something unique to humans may have been in the minds of early philosophers when they first considered the subject of thinking and intelligence, this does not really define the activity. Others maintain that a machine is thinking when it is performing activities that normally require thought when performed by human beings. Thus, adding 2+3 must be a form of thinking. To continue, some psychologists have defined intelligence in the following simple way: intelligence is what an intelligence test measures. In light of the preceding section on information systems, all that needs to be done is to feed enough information into an information system and to develop an appropriate query language, and the result is an intelligent machine. This line of reasoning also skirts a clear definition. Perhaps, it is a waste of time to worry about precise definitions, but the fact remains that computers are doing some amazing things - such as playing chess,

guiding robots, controlling space vehicles, recognizing patterns, proving theorems, and answering questions - and that these applications require much more than the conventional computer program. Richard Hamming, developer of the prestigious Hamming code for error detection and correction in computers, gives a definition of intelligent behavior that may be useful here:

> *The ability to act in subtle ways when presented with a class of situations that have not been exhaustively analyzed in advance, but which require rather different combinations of responses if the result in many specific cases is to be acceptable.*

Artificial Intelligence is an important subject because it may indicate the direction in which society is moving. Currently, machines are used for two reasons: (1) The job cannot be done by a human being, and (2) The job can be performed more economically by a machine. To this list, another reason must be added: some jobs are simply too dull to bed one by humans, and it is desirable from a social point of view to have such jobs done by machine. This requires a greater number of "intelligent" machines, since people seem to be finding more and more work they consider to be dull and routine.

Here are two items of before we get started with the talk:

> *Artificial general intelligence (AGI) is the intelligence of a machine that could successfully perform any intellectual task that a human being can. It is a primary goal of some AI research and is a common topic in science fiction and future studies. (Author unknown.)*

> *The term singularity is used as the hypothesis that the invention of artificial super intelligence (ASI) will abruptly trigger runaway technical growth, resulting in unfathomable change to human civilization. (Author unknown)*

It is possible to approach Artificial Intelligence from two points of view. Both approaches make use of programs and

programming techniques. The first approach is to investigate the general principles of intelligence. The second is to study human thought, in particular.

Those persons engaged in the investigation of the principles of intelligence are normally charged with the development of systems that appear to be intelligent. This activity is commonly regarded as artificial Intelligence, which incorporates both engineering and computer science components.

Those persons engaged in the study of human thought attempt to emulate human mental processes to a lesser or greater degree. This activity can be regarded as a form of "computer simulation," such that the elements of a relevant psychological theory are represented in a computer program. The objective of this approach is to generate psychological theories of human thought. The discipline is generally known as "Cognitive Science."

In reality, the differences between artificial intelligence and cognitive science tend to vary between "not so much" and "quite a lot" - depending upon the complexity of the underlying task. Most applications, as a matter of fact, contain elements from both approaches.

The Scope of AI

It is possible to zoom in on the scope of AI by focusing on the processes involved. At one extreme, the concentration is on the practicalities of doing AI programming, with an emphasis on symbolic programming languages and AI machines. In this context, AI can be regarded as a new way of doing programming. It necessarily follows that hardware/software systems with AI components have the potential for enhanced end-user effectiveness.

At the other extreme, AI could be regarded as the study of intelligent computation. This is a more grandiose and

encompassing focus with the objective of building a systematic and encompassing focus with the objective of building a systematic theory of intellectual processes - regardless of if they model human thought or not.

It would appear, therefore, that AI is more concerned with intelligence in general and less involved with human thought in particular. Thus, it may be contended that humans and computers are simply two options in the genus of information processing systems.

The Modern Era of Artificial Intelligence

The modern era of artificial intelligence effectively began with the summer conference at Dartmouth College in Hanover, New Hampshire in 1956. The key participants were Shannon from Bell Labs, Minsky from Harvard (later M.I.T.), McCarthy from Dartmouth (later M.I.T. and Stanford), and Simon from Carnegie Tech (renamed Carnegie Mellon). The key results from the conference were twofold:

- It legitimized the notion of AI and brought together a raft of piecemeal research activities.
- The name "Artificial Intelligence" was coined and the name more than anything had a profound influence of the future direction of artificial intelligence.

The stars of the conference were Simon, and his associate Allen Newell, who demonstrated the Logic Theorist - the first well-known reasoning program. They preferred the name, "Complex Information Processing," for the new fledging science of the artificial. In the end, Shannon and McCarthy won out with the zippy and provocative name of artificial intelligence. In all probability, the resulting

controversy surrounding the name artificial intelligence served to sustain a certain critical mass of academic interest in the subject - even during periods of sporadic activity and questionable results.

One of the disadvantages of the pioneering AI conference was the simple fact that an elite group of scientists was created that would effectively decide "what AI is and what AI isn't," and how to best achieve it. The end result was that AI became closely aligned with psychology and not with neurophysiology and to a lesser degree with electrical engineering. AI became a software science with the main objective of producing intelligent artifacts. In short, it became a closed group, and this effectively constrained the field for a large degree. In a sense, that is the way the field exists today.

In recent years, the direction of AI research has been altered somewhat by an apparent relationship with brain research and cognitive technology, which is known as the design of joint human-machine cognitive systems. Two obvious fallouts of the new direction are the well-known "Connection Machine," and the computer vision projects at the National Bureau of Standards in their United States. That information is somewhat out of date, but the history gives some insight into what AI is today and where it will be heading.

Early Work on the Concept of Artificial Intelligence

The history of AI essentially goes back to the philosophy of Plato, who wrote that. "All knowledge must be state able in explicit definitions which anyone could apply," thereby eliminating appeals to judgment and intuition. Plato's student Aristotle continued in this noble tradition in the development of the categorical syllogism, which plays an important part in modern logic.

The mathematician Leibnitz attempted to quantify all knowledge and reasoning through an exact algebraic system by which all objects are assigned a unique characteristic number. Using these characteristic numbers, therefore, rules for the combination of problems would be established and controversies could be resolved by calculation.

The underlying philosophical idea was conceptually simple: Reduce the whole of human knowledge into a single formal system. The notion of formal representation has become the basis of AI and cognitive science theories since it involves the reduction of the totality of human experience to a set of basic elements that can be glued together in various ways.

To sum up, the philosophical phenomenologists argue that it impossible to subject pure phenomena - i.e., mental acts which give meaning to the world - to formal analysis. Of course, AI people do not agree. They contend that "there is no ghost in the machine," and this is meant to imply that intelligence is a set of well-defined physical processes.

The discussion is reminiscent of the mind/brain controversy and it appears that the AI perspective is that "the mind is what the brain does." Of course, the phenomenologists would reply that the definition of mind exists beyond the physical neurons; it also incorporates the intangible concepts of what the neurons do.

Accordingly, strong AI is defined in the literature as the case wherein an appropriately programmed computer actually is a "mind." Weak AI, on the other hand, is the emulation of human intelligence, as we know it.

Intelligence and Intelligent Systems

There seems to be some value in the ongoing debate over the intelligence of AI artifacts. The term "artificial" in artificial

intelligence helps us out. One could therefore contend that intelligence is natural if it is biological and artificial otherwise. This conclusion skirts the controversy and frees intellectual energy for more purposeful activity.

The abstract notion of intelligence, therefore, is conceptualized, and natural and artificial intelligence serve as specific instances. The subjects of understanding and learning could be treated in a similar manner. The productive tasks of identifying the salient aspects of intelligence, understanding, and learning emerge as the combined goal of AI and cognitive science. For example, the concepts of representation and reasoning, to name only two of many, have been studied productively from both artificial and biological viewpoints. Software products that are currently available can be evaluated in the basis of how well they can support the basic AI technologies.

The key question then becomes: How well do natural and artificial systems, as discussed above, match up to the abstract notion of intelligence.

Cognitive Technology

Cognitive technology is the set of concepts and techniques for developing joint human-machine cognitive systems. People are obviously interested in cognitive systems because they are goal directed and employ self-knowledge of the environment to monitor, plan, and modify their actions in the pursuit of their goals. In a logical sense, joint human-machine systems can also be classed as being cognitive because of the availability of computational techniques for automating decisions and exercising operational control over physical processes and organizational activities.

Recent advances in heuristic techniques coupled with methods of knowledge representation and automated reasoning

have made it possible to couple human cognitive systems with artificial cognitive systems. Accordingly, joint systems in this case would necessarily have the following attributes:

- Be problem driven, rather than technology driven.
- Effective models of underlying processes are needed.
- Control of decision-making processes must be shared between human and artificial
- Clearly, cognitive technology represents a possible (if not probable) paradigm shift whereby the human self-view can and wake in the not-too-distant future.

Virtual Systems and Imagination

Methods for reasoning in expert and cognitive systems are well defined. Rules and representation effectively solve the problem. There appears to be a set of problems, however, that seem to evade such a simple solution as rules and representation.

A sophisticated model of a cognitive system must incorporate the capability of reasoning about itself or another cognitive system and about the computational facilities that provide the cognition. When a person, for example, is asked to reason about the feelings of the probable response of another person, set of rules is normally invoked to provide the desired response. If no rule set exists, then a virtual process is engaged that proceeds somewhat as follows:

- The object process is imagined.
- The neural inputs are "faked" and the brain responds in the usual manner.
- The result is observed exactly as though it had taken place.

Thus, a sort of simulation of a self-model is employed. This type of analysis might be invoked if someone were asked, for

example, how they would feel if they had just received the news they had contracted an incurable disease.

The process, described above, is essentially what an operating system does while controlling the execution of a *guest* operating system. Inputs and outputs are interpreted, but machine code is actually executed.

It necessarily follows those executable models, as suggested here, are as much a form of knowledge as are rules and facts.

But, Is It intelligent?

Bandying the issue even further, a sharp borderline between intelligent and non-intelligent behavior, in the abstract sense, probably does not exist. Nevertheless, some essential qualities might be the following:

- To respond to situations very flexibly.
- To take advantage of fortuitous circumstances.
- To make sense out of ambiguous or contradictory input.
- To recognize the relative importance of different elements of a situation.
- To find similarities between situations despite differences which may separate them.
- To draw distinctions between situations despite similarities which may link them.
- To synthesize new concepts by taking old concepts and putting three together in new ways.
- To come up with ideas that they are novel.

Viewed in this manner, intelligence is a form of computation. Effective intelligence then is a process (perhaps a computer program) and an appropriate machine in which to execute the process.

Systems Concepts and AI

An interesting viewpoint concerns the evolution of data processing has emerged from the AI business. The task of designing a rule base and an associate fact base is somewhat analogous to designing an information system. Moreover, both kinds of systems appear to evolve in a similar manner. For this analysis, it should be assumed that statements and computational processes (i.e., modules) are the same (or synonymous).

A sensory stimulation is associated with a statement. (Incidentally, this concept is known as associations, wherein a sensation is associated with an idea, and that idea leads to another idea, and so forth. This theory originated with Aristotle and was pursued by Hobbs, Locke, and Mill.). The associations reverberate through the system of statements, whereby a result is finally achieved. The output can be viewed as a prediction. Moreover, the system operates according to some form of internal laws - such as the laws of mathematics.

When a prediction fails, continuing with the statement analogy, we question the validity of the set of statements. Revisions are normally in order. Since a direct correlation is usually possible between the stimuli and peripheral statements, these statements are preserved from revision. Other statements must bear the brunt of change. The other statements, however, can be regarded as the Starting Point, since they result from internal laws. The end result that a priority judgment. Is necessary: change the peripheral statements of change the frozen middle. The priorities of course are in conflict and the preference commonly goes to the revision that disturbs the system the least.

Effectively, incremental changes are made to the system until a total revision is necessary - i.e., a rewrite of the internal including the Laws, the frozen middle, and the peripheral

statements. As a total concept, major revisions serve to simplify a system. It necessarily follows that some attention should be given to systems evolution as a predictive technique.

END OF CHAPTER SEVEN

Chapter 8

Thinking

An academic named John Dewey specialized in thinking, at least for a while anyway, and wrote the popular book entitled *How We Think* that was translated into several languages. At first, the emphasis of thinking was to solve problems and consisted of the scientific method, reflective thinking, and the role of the educator in public life; it consisted of reading, writing, and arithmetic and subsequently the academic subjects of history, mathematics, and chemistry. Dewey's work was organized around some aspect of educational learning, associated with the Laboratory School at the University of Chicago. Dewey defined reflective thought as the active, persistent, and appropriate examination of beliefs of the information that is available. Thoughts are a continuum of several beliefs. Dewey says that the scientific method of behavior involves deferring to make judgment until sufficient data is obtained and testing, or verification, is done. Dewey has developed five steps, relevant to AI, that constitute reflective thought. (1) Development of the uncertainty requiring the reflective thought. (2) Defining the problem to be solved. (3) Collection of data and other factors relevant to the problem. (4) Construction of suitable hypotheses. (5) Development of methods to solve the problem.

One of the leading persons in developing conceptual ideas about intelligent computers was an English scientist name Alan Turing, the mathematician that broke the German cryptographic code that helped the U.S., Great Britain, and other countries - called

the Allies - win World War II," continued Matt. Alan Turing was a forward looking individual and was involved with several academic pursuits. One of the most popular was the *Turing Test* for determining whether a computer and its program were intelligent or not. On the surface, it seems pretty simple minded, but it gave people something to think about; I am supposed to be looking into thinking. He proposed that if a person gave a question to a person and a computer, and could not tell if the written answer was from the computer or a person then we could say that the computer was thinking, that is exhibiting intelligence, that is artificial intelligence. It seems to be a little simple minded, but it makes some sense. Of course, there is no mention of how the computer would do it.

Relevant concepts on thinking that would serve as an aid to someone desiring to learn something about thinking.

1. Every involving thought has to go through the mind, person or computer.
2. Everything we say goes through our heads.
3. Thinking usually involves matters not directly perceived.
4. Thought involves belief in something accepted or believed.
5. Thoughts implied by certain beliefs are the source of additional beliefs.
6. Thoughts caused by events remind a person of related consequences.
7. Curiosities that cause senses are achieved through thinking.

Why is thinking of concern to the development of AI systems? That is because AI thinking precludes AI behavior.

END OF CHAPTER EIGHT

Chapter 9

Learning

The term *learning* in an Artificial Intelligence context is the form of activity that results in purposeful behavior. In short, the AI software learns what to do to solve a problem in very much the same way it is done in the minds of humans. In AI systems, the AI software and human thinking work together to resolve a specific class of problems."

In AI, what is learned? Is it the method, the solution, or both.

Machine intelligence and learning reflects a computer system that exhibits behavior that would be classified as intelligent if performed by a human agent. That computer system would be known as an intelligent agent, and that field of study is termed *artificial intelligence*. Some researchers refer to AI, as it is usually called, as that which hasn't been done yet, even though that term doesn't seem to make sense in today's computer world. That would imply that an intelligent agent should, in fact, perform a certain amount of learning.

Some researchers feel as though AI can be achieved when human intelligence can be described in sufficient detail that a machine can be made to simulate it. One of the early artificial intelligent applications is computer chess, wherein the program emulates a human chess player. Some AI applications can learn how to play a game and win by learning how to operate without human intervention.

In general, AI researchers feel that any problem that can be

solved by a human can be developed as an AI application, and the challenger is to have the AI program learn by itself how to play. That is why machine learning is one of the most important concepts in artificial intelligence.

Other disciplines can be related to machine intelligence, such as advanced statistical techniques, which is termed *deep learning*. There are several topics related to machine intelligence and learning:

- Reasoning and problem solving
- Knowledge representation
- Knowledge engineering
- Qualification
- Natural language
- Creativity
- General intelligence

AI is relevant to any task that can demonstrate intelligence. As before, when a term (or technique) reaches mainstream, it is no longer considered artificial intelligence. This is the *AI effect*. Here are some worthy applications: Healthcare, Automotive, Finance, and Video Games. There are two basic questions:

- Is artificial general intelligence possible? Can a machine solve any problem that a human can solve?
- Are intelligent machines dangerous?"

Machine learning is a complicated subject in artificial intelligence.

END OF CHAPTER NINE

Chapter 10

The Advisor Model

An expert-system is characterized by a database of facts and a logical engine for exploiting the facts in a purposeful manner. The logical engine takes good advantage of the available information so that the output of the system is considerably more than what would ordinarily be implied strictly by the input. The data and the inference programs give the system an air of being an expert.

One of the most illustrative examples of an expert system is the well-known wine advisor. The user enters a brief description of an entrée, such as chicken with cream sauce, and the system yields a small wine selection, such as Chardonnay or Montrachet – depending of course upon the current contents of the wine cellar. Naturally, there is more to the story. The entrée has certain characteristics that are determined by a database search. The characteristics are then used by an inference mechanism during a "run" of the system to generate the attributes of a suitable wine, such as color, taste, and strength (i.e., light, heavy, etc.) In a second database operation, the wine attributes are mapped into a "variety" of wine. Finally, a selection is made from the database that represents the contents of the wine cellar.

Similar advisors have been constructed in the financial community and in the nuclear industry – to name only a few. In an investment advisor, for example, an investor's profile is mapped into a well-defined, but much larger, set of investment

characteristics using a database operation. Again, an inference mechanism is used; in this case, it generates a portfolio structure that serves as input to a database search operation yielding a set of investment options. A typical portfolio structure might be: 60% equities, 35% bonds, and 5% gold. The output set would be a small selection of instruments in selected currencies.

In the nuclear industry, a collection of physical events occurring in an emergency situation would be translated by an expert system into a set of instructions to an operator. As before, the processing sequence includes an initial database search operation, an inference procedure, and then a final database selection to generate the needed instructions. The examples typify the "advisor model."

The *advisor model* is characterized by the following set of actions occurring in essentially the order given:

- A relatively small set of input values are entered into the system via and end user interaction or as the result of a physical event.
- The input values are expanded into the particulars for a certain problem domain via a database search process.
- The particulars are mapped into a set of descriptors via an inference mechanism that can take the form of a set of if-then rules and a logical engine, a neural network, a pattern matching operation, a mathematical function, or possibly logic program.
- The descriptors are used as key values in a subsequent database search operation to yield the results for that problem.

Clearly, the salient properties of the advisor model are the initial database operation, the inference (or mapping) process, and the final database search. The two database operations are

obviously to different structures. Each of these components can be replicated, as required by a particular problem and its problem solver, but the general structure of the model is evident without necessarily going into more detail. In fact, most expert systems operate in precisely this manner, even though the emphasis has conventionally been on the rules and knowledge-based parts of the model.

The advisor model is independent of the mode of representation; this fact is tacitly understood and is inherent in the notion of a model. Representation is dependent upon a specific instantiation of the model. If the cells in a spreadsheet application are the source of information, then the retrieval and inference operations are related to that framework. If the inference processes are implemented as LISP or Prolog programs, then the particulars and descriptors are constrained by the corresponding ad hoc data structures. To sum up, the method of representation is determined by the underlying problem domain and computational framework. This explains to a limited degree why a lot more "advisors" are not floating around in the industrial, academic, and governmental communities. The ultimate objective is clear: to make the benefits of the advisor model available in a general-purpose format.

Another consideration that complicates the picture is the existence of uncertainty in most areas dealings with human affairs. Uncertainty has been the Achilles heel of the artificial intelligence community, since a practical system for problem solving should take the uncertainty of information or the uncertainty of the reasoning process into consideration. Naturally, we would like to have the mechanisms in place for dealing effectively with both forms of uncertainty, even though all of the features may not be needed for a particular problem situation. It is indeed one thing to map an entrée of beef into a wine descriptor in a deterministic sense, such as:

beef -> strong red meat -> dry heavy red wine -> ...

and quite another to introduce the notion of uncertainty, as in:

*small lump -> (((healthy, tests))),0.6),((healthy, tests, sick),0.4))
->... -> ((home, office).0.56)*

The notation in the latter example is intended to demonstrate that a *small lump* is mapped into the conditions of *healthy* or *tests* with a certainty of 0.6 and into the conditions of *healthy* or *tests* or *sick* with a certainty factor of 0.4. Total comprehension here is not expected, but is given to indicate that there are different ways to use AI.

In the wine and medical examples, there is relatively little complication in the advisory process, and gives us the opportunity of focusing on the initial database operation. The process represents a translation between the external world of the end user to the internal world of the automated problem solver In the wine advisor, for example, the indicator of beef is mapped into a set of attributes with which AI model can work, as in:

beef -> strong red meat

or in its generic form:

indicator -> meaning

The notion of an indicator is well-established in organized society in the form of symbols, variables, synonyms, and abbreviations. It is commonplace to have an indicator, such as *successful,* that is associated with a substantial set of characteristics, such as *aggressive, confident, intelligent, and focused.* Indicators can have different meanings to different people' however, which preserves the role of the expert. In the

absence of uncertainly, and indicator has a definite meaning. In the presence of uncertainty, an indicator *might* mean something, and the degree to which an expert believes the assertion is known as its certainty value, or as its the certainty factor, as it is sometimes called. Thus, the indicator in the presence of uncertainty is a clue to the validity of an assertion; it becomes evidence when combined with its meaning and a certainty value, or in its generic form:

Evidence: *indicator -> meaning with certainly value.*

An obvious consideration applies to the notion of evidence. As in common everyday usage, a fact "is entered as evidence" only when it is relevant to the case under consideration. An analogous caveat applies to the methods contained here, and a means is presented for constraining the relevancy of information.

END OF CHAPTER TEN

Chapter 11

Conversational Programs

Alan Turing is an English mathematician whose work on cryptography was a major contribution to the victory in World War II. Almost single handedly, he developed a machine that could solve the German's cryptographic enigma code that was used to win the war for the allies. He is also very well known for the development of basic concepts in mathematics and computation. Alan Turing received his PhD from Princeton University.

Dr. Turing is best known for a variety of intellectual concepts, but the most well-known is the Turing Test for artificial intelligence. The fundamental idea is based on the idea that a computer exhibits intelligence when it can answer questions, and a person can or cannot tell that it is a computer answering the questions. The basic result is that if a computer is mistaken for a human more than 30% of time during a set of keyboard interactions, then it is exhibiting intelligence. The test works as follows. There are three computers connected together and two people and one supposedly intelligent program. One person asks the questions and the other two computers answer the question. If the program can answer the questions and it is undetectable that is the computer, then that program exhibits intelligence. Turing also gives conditions for conducting the "Turing test," but that is the basic idea behind the test.

There is another AI-type program named ELIZA developed by Joseph Weizenbaum at MIT in 1966. The on-line program uses the input to determine the response, as in non-directive counseling in Psychology. ELIZA is on-line and here is no secret about how it operates. Yet, so many people begin to believe it, that many universities - including MIT - had to take it offline.

The user person just says things to ELIZA and ELIZA responds just like an ordinary person would, using the information in the person's input. The ELIZA program is not that simple. It has to stay on the beam, so to speak. Most people don't think of that when they think of Artificial Intelligence, but it would seem to have intelligence.

There is an Experimental Scientist from Hungary that has looked into the subject of conversational programs and has written a book on the subject. I'll give the reference later. The program, ELIZA, contains a collection of patterns and a collection of answers corresponding to the patterns. There is another program, like DOCTOR. There is a third system named PARRY. It is interesting when they put DOCTOR and PARRY together. The PARRY program also is known to deceive practicing psychiatrists.

The reference is: Mérö, László, *Ways of Thinking: The Limits of Rational Thought and Artificial Intelligence,* New Jersey: World Scientific Publishing Co., 1990. There is a movie on Alan Turing titled *The Imitation Game.*

END OF CHAPTER ELEVEN

Chapter 12

How We Think

There is not a lot of literature on precisely what thinking is. An academic named John Dewey specialized in it, at least for a while anyway, and wrote the popular book entitled How We Think. At first, the emphasis of thinking was to solve problems and consisted of the scientific method, reflective thinking, and the role of the educator in public life; it consisted of reading, writing, and arithmetic and subsequently the academic subjects of history, mathematics, and chemistry. Dewey defined reflective thought as the active, persistent, and appropriate examination of beliefs of the information that is available. Thoughts are a continuum of several beliefs. Dewey says that the scientific method of behavior involves deferring to make judgment until sufficient data is obtained and testing, or verification, is done. Dewey has developed five steps, relevant to AI, that constitute reflective thought. (1) Development of the uncertainty requiring the reflective thought. (2) Defining the problem to be solved. (3) Collection of data and other factors relevant to the problem. (4) Construction of suitable hypotheses. (5) Development of methods to solve the problem.

The term learning in an Artificial Intelligence context is the form of activity that results in purposeful behavior. In short, the AI software learns what to do to solve a problem in very much the same way it is done in the minds of humans. In AI systems, the AI software and human thinking work together to resolve a

specific class of problems. In AI, what is learned? Is it the method, solution, or both.

At this point, it seems like nothing is learned. Some smart person designs a problem and writes a computer program to solve the problem, and it is called Artificial Intelligence and it supposed to be intelligent. That's the problem with this problem, information is scattered all over the place and who knows what it means. Here is a example of the current situation, as a whole. "Go to a well-known retail store that sells just about everything including food and computers, looking for a cartridge for your printer, and there is a kid working there who says 'we have computers and artificial intelligence'. He really should have said software, but how was he to know what artificial intelligence is unless there is some reliable place where he should go for information. You can see what you are up against. We are finished with thinking and are looking for some small applications to use as examples.

END OF CHAPTER TWELVE

Chapter 13

A Quick Look at ChatGPT

ChatGPT is called a form of *Generative AI*, like everyone reading about it knows what Generative AI is. Here is a list all of those new made-up names. At least the ones that are easy to run across. In a system of this type, there seem to be three entities: a foundation of information in various forms, an application model that accesses the data, that can be any form imaginable, and a sophisticated front that generates the information in the form the user requests it. All three entities are complicated. In my mind, ChatGPT is a system that generates information for the end user. It is analogous to the big computer, years ago called a mainframe, that initially performed a task for the end user."

Here is a list of ChatGPT definitions:

ChatGPT stands for Generative Pre-trained Transform.
ChatGPT is analogous to the results produced on customer service websites - ask questions and get answers.
ChatGPT is an AI program called a chatbot that employs natural language processing to create a humanlike conversational dialogue.

The language model responds to questions and can create various forms of written context, such as articles, messages, emails, computer programs, and media posts. Maybe even artwork.

The natural language processing includes machine translation, information retrieval, sentence analysis, and information extraction.

ChatGPT is a kind of generative AI that involves entering prompts and receiving humanlike images, test, or video.

ChatGPT is similar it automated results and services that are found on customer service websites: ask questions and receive clarification.

ChatGPT is trained with reinforcement learning through human feedback and reward systems that analyze and rank the responses. This is machine learning.

ChatGPT is a product of ca company named OpenAI.

ChatGPT uses algorithms to find patterns within data sequences.

ChatGPT uses deep learning to produce text with the use of neural networks.

ChatGPT produces text, next word, sentences, or paragraphs based on some learning formulation.

Human trainers provide conversations to test and rank results.

ChatGPT has been used to code computer programs, compose music, writ emails, summarize articles, write social memos, make titles, solve math problems s, create articles, reword existing content, make product descriptions, play games, do job searches, and ask questions like in ELIZA.

ChatGPT has limitations does have limitations, such as the complexity of human language, use up to date data, poor on analytics, doesn't get the subtleties of modern language, and has trouble focusing on what data is relevant to a problem.

ChatGPT has some ethics problems such as privacy, security, cheating in education, and cheating at work.

ChatGPT has problems of plagiarism, bias, fireplaces, human jibs, and cost.

Someone has said that ChatGPT is a good place to start learning what artificial intelligence is all about.

END OF CHAPTER THIRTEEN

Chapter 14

A Very Quick Summary of Artificial Intelligence

In the next couple of chapters and the final report, here is a quick summary of what you are going to have to know to understand what is going on in the AI world:

Machine Learning

Machine learning is area of study in Artificial Intelligence that is. Usually divided into two categories: Supervisor learning and Unsupervised learning. It is usually considered when designing large scale AI systems. It is normally referred to as an enabler.

Cognitive Computing

Cognitive computing is a technique to develop computers that behave like the human mind through the AI techniques such as expert systems, neural networks, and machine learning. This statement has been tagged for a rewrite.

Types of AI

Types of AI includes most concepts that exist within the AI domain, such as: weak AI, strong AI, deep learning, machine learning, reactive machines, spatial computing, generative AI,

facial recognition, chat bots, synthetic data, contrastive learning, and so on. The is a complex section.

Watson System

Watson is an IBM system that provides a conversational AI system that can service in a large variety of disciplines. It typically is integrated into a larger system as services as an interactive assisted designed to handle large databases and other information domains. Practically every element of existing AI technology is provided with Watson.

END OF CHAPTER FOURTEEN

Chapter 15

Machine Learning

Artificial Intelligence Specifics

It is the study and development of techniques for making machines that people would class as being smart, especially computer systems and programs. It involves the study of human intelligence, but it is not limited to what we would class as human activity. Thus, the subject of Artificial Intelligence defines the concept of intelligence for its own purpose. Like saying a checkers program is intelligent.

At this point in its development, there are no systems that are more intelligent than humans. There are programs that do very complicated things, however, and they are under the control of other computers or human beings.

The goals of Artificial Intelligence are the development of machines - usually computers - that can think and act like human beings and incorporate thinking and acting accordingly.

Theoretically, the scope of Artificial Intelligence ranges from simple programs to systems that are more intelligent than humans. Many people believe that Artificial Intelligent machines will take over the civilized world, and just as many, perhaps more, think that is not true.

Machine Learning Specifics

There are two types of Artificial Intelligence: weak AI and strong AI. *Weak AI is* sometimes referred to as *Narrow AI,* since people like fancy names and also *Artificial Narrow Intelligence* (ANI), that are designed to perform specific small tasks. Some ANI systems are exceedingly simple like vacuum cleaners that can roam around the living room, and a little more complicated self-driving cars and IBM's Watson.

Strong AI is a group of AI's, including Artificial General Intelligence (AGI) or Artificial Super Intelligent (ASI) with intelligence equal to humans. It could do anything, such as solve problems and plan for the future. It would have super intelligence. Strong idea is a good idea, to some degree, but doesn't exist today.

This opens the door to deep learning and machine learning. Since the two names are used interchangeably, it is prudent to denote the difference between them. Machine learning is a sub area of Artificial Intelligence, and deep learning is a sub area of machine learning.

The types of learning differ in how the algorithm learns. Non-deep learning uses human experts to analyze data while deep learning does it by itself.

Deep learning employs neural networks and is generally considered mor of what Artificial Intelligence should be. Neural networks is with Quantum Computing as two of the most complicated subject in advance computer technology.

Generative models take raw data and generate useful information. Generative models take raw data and generate useful information. Generate models are not new and are what the subject of statistics is designed to do. The modern Artificial

Intelligence method use information to select data for display or other forms of computation.

Artificial Intelligence Applications

Most Artificial Intelligence writers summarize into 5 categories. They are: Speech Recognition, Customer Service, Computer Vision, Recommendation Engines, and Automated Stock Trading. Here is a brief summary of those categories.

Speech: Uses natural language processing to analyze human speech and generate it in a written form.

Customer: Called virtual agents that respond to customer queries, that may be questions, products, and sizes through customer engagement. Referred to as virtual agents or voice assistants.

Vision: Used to derive useful information from digital images and videos. Used in healthcare, law enforcement, or design by computer.

Recommendation: Use past consumption to discover data trends to improve various business goals.

Automated: Used to optimize stock portfolios and facilitate automated stock trading.

There you have it.

END OF CHAPTER FIFTEEN

Chapter 16

Cognitive Computing

The fundamental stating point of modern AI is the nebulous subject generally known as cognitive computing. Here is a list of cognitive subjects:

 Conventional computing and cognitive computing
 Machine conversation
 Compute vision
 Computer speaking
 Empathy from computers
 Discovery and understanding of unstructured content
 (Watson subject)
 Cognitive systems (Watson subject)

 On the surface the list is subject are straight forward subject, but they are exceedingly complicated.

Conventional computing and cognitive computing

 Convention computing is, as practically everyone knows, is a pre-planned set of steps to provide a per-planned result. With convention, applications are conceived, designed, programmed or coded, tested, and then put into operation through a person-oriented scenario. The only intelligence is in the mind of the person's involved. The objectives are identified beforehand, and additionally, computing elements are designed to interact of a pre-defined manner.

With cognitive computing, elements are defined and combined to execute in an intelligent manner without human intervention. The computing elements may be defined and prepared or exist in the outside operational environment.

Machine Conversation

Machine conversation is a defined collection of natural language processing and machine learning elements. In its most simple form it is two conversational programs that talk to one another, such as ELIZA and ELIZA or DOCTOR and DOCTOR, mentioned previously in this report. Intelligent machine conversation involves two compatible intelligent entities interactions with one another.

Computer Speaking

The ability to initiate computer speech is a companion to speech recognition. It is usually combined with speech recognition and cognitive computing. A simple example is text reading and speaking. When combined with other cognitive elements, it is in the AI domain.

Empathy from Computers

There are two kinds of computer empathy: human empathy for computer technology and its various difficulties, and the empathy of computers to humans. Computer can sense emotion if defined beforehand. As yet, it has not been considered in the area of Artificial Intelligence.

Cognitive Systems

The discovery and understanding of unstructured information and cognitive systems is generally combined under the title Generative Models.

A generative model is a deep learning facility that can take raw data, such as database information, speech, visual information, and other event data and draw relevant conclusions.

END OF CHAPTER SIXTEEN

Chapter 17

Artificial Intelligence Final Report
(Duplications for continuity)

Summary

Overview

The question, "Can a machine think?" is one that has been debated for some time now and the question is no likely to be answered in this book. However, the subject is fruitful when considering "what a computer can do."

There are various opinions on the subject. Some say that thinking is an activity that is peculiar to human beings. Accordingly, machines cannot think. Although thought as something unique to humans may have been in the minds of early philosophers when they first considered the subject of thinking and intelligence, this does not really define the activity. Others maintain that a machine is thinking when it is performing activities that normally require thought when performed by human beings. Thus, adding 2+3 must be a form of thinking. To continue, some psychologists have defined intelligence in the following simple way: intelligence is what an intelligence test measures. In light of the preceding section on information systems, all that needs to be done is to feed enough information into an information system and to develop an appropriate query language, and the result is an intelligent machine. This line of

reasoning also skirts a clear definition. Perhaps, it is a waste of time to worry about precise definitions, but the fact remains that computers are doing some amazing things - such as playing chess, guiding robots, controlling space vehicles, recognizing patterns, proving theorems, and answering questions - and that these applications require much more than the conventional computer program. Richard Hamming, developer of the prestigious Hamming code for error detection and correction in computers, gives a definition of intelligent behavior that may be useful here:

The ability to act in subtle ways when presented with a class of situations that have not been exhaustively analyzed in advance, but which require rather different combinations of responses if the result in many specific cases is to be acceptable.

Artificial Intelligence is an important subject because it may indicate the direction in which society is moving. Currently, machines are used for two reasons: (1) The job cannot be done by a human being, and (2) The job can be performed more economically by a machine. To this list, another reason must be added: some jobs are simply too dull to bed one by humans, and it is desirable from a social point of view to have such jobs done by machine. This requires a greater number of "intelligent" machines, since people seem to be finding more and more work they consider to be dull and routine.

Here are two items of before we get started with the talk:

Artificial general intelligence (AGI) is the intelligence of a machine that could successfully perform any intellectual task that a human being can. It is a primary goal of some AI research and is a common topic in science fiction and future studies. (Author unknown.)

The term singularity is used as the hypothesis that the invention of artificial super intelligence (ASI) will abruptly trigger runaway

technical growth, resulting in unfathomable change to human civilization. (Author unknown)

It is possible to approach Artificial Intelligence from two points of view. Both approaches make use of programs and programming techniques. The first approach is to investigate the general principles of intelligence. The second is to study human thought, in particular.

Those persons engaged in the investigation of the principles of intelligence are normally charged with the development of systems that appear to be intelligent. This activity is commonly regarded as "artificial intelligence," which incorporates both engineering and computer science components.

Those persons engaged in the study of human thought attempt to emulate human mental processes to a lesser or greater degree. This activity can be regarded as a form of "computer simulation," such that the elements of a relevant psychological theory are represented in a computer program. The objective of this approach is to generate psychological theories of human thought. The discipline is generally known as "Cognitive Science."

In reality, the differences between artificial intelligence and cognitive science tend to vary between "not so much" and "quite a lot" - depending upon the complexity of the underlying task. Most applications, as a matter of fact, contain elements from both approaches.

The Scope of AI

It is possible to zoom in on the scope of AI by focusing on the processes involved. At one extreme, the concentration is on the practicalities of doing AI programming, with an emphasis on symbolic programming languages and AI machines. In this

context, AI can be regarded as a new way of doing programming. It necessarily follows that hardware/software systems with AI components have the potential for enhanced end-user effectiveness.

At the other extreme, AI could be regarded as the study of intelligent computation. This is a more grandiose and encompassing focus with the objective of building a systematic and encompassing focus with the objective of building a systematic theory of intellectual processes - regardless of if they model human thought or not.

It would appear, therefore, that AI is more concerned with intelligence in general and less involved with human thought in particular. Thus, it may be contended that humans and computers are simply two options in the genus of information processing systems.

The Modern Era of Artificial Intelligence

The modern era of artificial intelligence effectively began with the summer conference at Dartmouth College in Hanover, New Hampshire in 1956. The key participants were Shannon from Bell Labs, Minsky from Harvard (later M.I.T.), McCarthy from Dartmouth (later M.I.T. and Stanford), and Simon from Carnegie Tech (renamed Carnegie Mellon). The key results from the conference were twofold:

It legitimized the notion of AI and brought together a raft of piecemeal research activities.

The name Artificial Intelligence was coined and the name more than anything had a profound influence of the future direction of artificial intelligence.

The stars of the conference were Simon, and his associate Allen Newell, who demonstrated the Logic Theorist - the first

well-known reasoning program. They preferred the name, "Complex Information Processing," for the new fledging science of the artificial. In the end, Shannon and McCarthy won out with the zippy and provocative name, "artificial intelligence." In all probability, the resulting controversy surrounding the name artificial intelligence served to sustain a certain critical mass of academic interest in the subject - even during periods of sporadic activity and questionable results.

One of the disadvantages of the pioneering AI conference was the simple fact that an elite group of scientists was created that would effectively decide "what AI is and what AI isn't," and how to best achieve it. The end result was that AI became closely aligned with psychology and not with neurophysiology and to a lesser degree with electrical engineering. AI became a software science with the main objective of producing intelligent artifacts. In short, it became a closed group, and this effectively constrained the field for a large degree. In a sense, that is the way the field exists today.

In recent years, the direction of AI research has been altered somewhat by an apparent relationship with brain research and cognitive technology, which is known as the design of joint human-machine cognitive systems. Two obvious fallouts of the new direction are the well-known "Connection Machine," and the computer vision projects at the National Bureau of Standards in their United States. That information is somewhat out of date, but the history gives some insight into what AI is today and where it will be heading.

Early Work on the Concept of Artificial Intelligence

The history of AI essentially goes back to the philosophy of Plato, who wrote that. "All knowledge must be state able in explicit

definitions which anyone could apply," thereby eliminating appeals to judgment and intuition. Plato's student Aristotle continued in this noble tradition in the development of the categorical syllogism, which plays an important part in modern logic.

The mathematician Leibnitz attempted to quantify all knowledge and reasoning through an exact algebraic system by which all objects are assigned a unique characteristic number. Using these characteristic numbers, therefore, rules for the combination of problems would be established and controversies could be resolved by calculation.

The underlying philosophical idea was conceptually simple: Reduce the whole of human knowledge into a single formal system. The notion of formal representation has become the basis of AI and cognitive science theories since it involves the reduction of the totality of human experience to a set of basic elements that can be glued together in various ways.

To sum up, the philosophical phenomenologists argue that it impossible to subject pure phenomena - i.e., mental acts which give meaning to the world - to formal analysis. Of course, AI people do not agree. They contend that "there is no ghost in the machine," and this is meant to imply that intelligence is a set of well-defined physical processes.

The discussion is reminiscent of the mind/brain controversy and it appears that the AI perspective is that "the mind is what the brain does." Of course, the phenomenologists would reply that the definition of mind exists beyond the physical neurons; it also incorporates the intangible concepts of what the neurons do.

Accordingly, strong AI is defined in the literature as the case wherein an appropriately programmed computer actually is a "mind." Weak AI, on the other hand, is the emulation of human intelligence, as we know it.

Intelligence and Intelligent Systems

There seems to be some value in the ongoing debate over the intelligence of AI artifacts. The term "artificial" in artificial intelligence helps us out. One could therefore contend that intelligence is natural if it is biological and artificial otherwise. This conclusion skirts the controversy and frees intellectual energy for more purposeful activity.

The abstract notion of intelligence, therefore, is conceptualized, and natural and artificial intelligence serve as specific instances. The subjects of understanding and learning could be treated in a similar manner. The productive tasks of identifying the salient aspects of intelligence, understanding, and learning emerge as the combined goal of AI and cognitive science. For example, the concepts of representation and reasoning, to name only two of many, have been studied productively from both artificial and biological viewpoints. Software products that are currently available can be evaluated in the basis of how well they can support the basic AI technologies.

The key question then becomes: How well do natural and artificial systems, as discussed above, match up to the abstract notion of intelligence.

Cognitive Technology

Cognitive technology is the set of concepts and techniques for developing joint human-machine cognitive systems. People are obviously interested in cognitive systems because they are goal directed and employ self-knowledge of the environment to monitor, plan, and modify their actions in the pursuit of their goals. In a logical sense, joint human-machine systems can also be classed as being cognitive because of the availability of computational techniques for automating decisions and exercising operational control over physical processes and organizational activities.

Recent advances in heuristic techniques coupled with methods of knowledge representation and automated reasoning have made it possible to couple human cognitive systems with artificial cognitive systems. Accordingly, joint systems in this case would necessarily have the following attributes:

- Be problem driven, rather than technology driven
- Effective models of underlying processes are needed.
- Control of decision-making processes must be shared between human and artificial components.

Clearly, cognitive technology represents a possible (if not probable) paradigm shift whereby the human self-view can and wake in the not-too-distant future.

Virtual Systems and Imagination

Methods for reasoning in expert and cognitive systems are well defined. Rules and representation effectively solve the problem. There appears to be a set of problems, however, that seem to evade such a simple solution as rules and representation.

A sophisticated model of a cognitive system must incorporate the capability of reasoning about itself or another cognitive system and about the computational facilities that provide the cognition. When a person, for example, is asked to reason about the feelings of the probable response of another person, set of rules is normally invoked to provide the desired response. If no rule set exists, then a virtual process is engaged that proceeds somewhat as follows:

- The object process is imagined.
- The neural inputs are "faked" and the brain responds in the usual manner.
- The result is observed exactly as though it had taken place.

Thus, a sort of simulation of a self-model is employed. This type of analysis might be invoked if someone were asked, for example, how they would feel if they had just received the news they had contracted an incurable disease.

The process, described above, is essentially what an operating system does while controlling the execution of a "guest" operating system. Inputs and outputs are interpreted, but machine code is actually executed.

It necessarily follows those executable models, as suggested here, are as much a form of knowledge as are rules and facts.

Is It intelligent?

Bandying the issue even further, a sharp borderline between intelligent and non-intelligent behavior, in the abstract sense, probably does not exist. Nevertheless, some essential qualities might be the following:

- To respond to situations very flexibly.
- To take advantage of fortuitous circumstances.
- To make sense out of ambiguous or contradictory messages.
- To recognize the relative importance of different elements of a situation.
- To find similarities between situations despite differences which may separate them.
- To draw distinctions between situations despite similarities which may link them.
- To synthesize new concepts by taking old concepts and putting three together in new ways.
- To come up with ideas that they are novel.

Viewed in this manner, intelligence is a form of computation. Effective intelligence then is a process (perhaps a computer

program) and an appropriate machine in which to execute the process.

Systems Concepts and AI

An interesting viewpoint concerns the evolution of data processing has emerged from the AI business. The task of designing a rule base and an associate fact base is somewhat analogous to designing an information system. Moreover, both kinds of systems appear to evolve in a similar manner. For this analysis, it should be assumed that statements and computational processes (i.e., modules) are the same (or synonymous).

A sensory stimulation is associated with a statement. (Incidentally, this concept is known as associations, wherein a sensation is associated with an idea, and that idea leads to another idea, and so forth. This theory originated with Aristotle and was pursued by Hobbs, Locke, and Mill.). The associations reverberate through the system of statements, whereby a result is finally achieved. The output can be viewed as a prediction. Moreover, the system operates according to some form of internal laws - such as the laws of mathematics.

When a prediction fails, continuing with the statement analogy, we question the validity of the set of statements. Revisions are normally in order. Since a direct correlation is usually possible between the stimuli and peripheral statements, these statements are preserved from revision. Other statements must bear the brunt of change. The other statements, however, can be regarded as the "frozen middle,' since they result from internal laws. The end result that a priority judgment. Is necessary: change the peripheral statements of change the frozen middle. The priorities of course are in conflict and the preference commonly goes to the revision that disturbs the system the least.

Effectively, incremental changes are made to the system until a total revision is necessary - i.e., a rewrite of the internal including the rules, the frozen middle, and the peripheral statements. As a total concept, major revisions serve to simplify a system. It necessarily follows that some attention should be given to systems evolution as a predictive technique.

End of Summary

Detailed Supplement

Thinking

An academic named John Dewey specialized in it, at least for a while anyway, and wrote the popular book entitled How We Think. At first, the emphasis of thinking was to solve problems and consisted of the scientific method, reflective thinking, and the role of the educator in public life; it consisted of reading, writing, and arithmetic and subsequently the academic subjects of history, mathematics, and chemistry. Dewey defined reflective thought as the active, persistent, and appropriate examination of beliefs of the information that is available. Thoughts are a continuum of several beliefs. Dewey says that the scientific method of behavior involves deferring to make judgment until sufficient data is obtained and testing, or verification, is done. Dewey has developed five steps, relevant to AI, that constitute reflective thought. (1) Development of the uncertainty requiring the reflective thought. (2) Defining the problem to be solved. (3) Collection of data and other factors relevant to the problem. (4)

Construction of suitable hypotheses. (5) Development of methods to solve the problem.

The term <u>learning</u> in an Artificial Intelligence context is the form of activity that results in purposeful behavior. In short, the AI software learns what to do to solve a problem in very much the same way it is done in the minds of humans. In AI systems, the AI software and human thinking work together to resolve a specific class of problems. In AI, what is learned? Is it the method, solution, or both.

At this point, it seems like nothing is learned. Some smart person designs a problem and writes a computer program to solve the problem, and it is called Artificial Intelligence and it supposed to be intelligent. That's the problem with this problem, information is scattered all over the place and who knows what it means. I went to a well-known retail store that sells just about everything including food and computers. I was looking for a cartridge for our printer, and this kid working there says we have computers and artificial intelligence. He really should have said software, but how was he to know what artificial intelligence is unless there is some reliable place where he should go for information. You can see what you are up against. We are finished with thinking and are looking for some small applications to use as examples.

A Quick Look at ChatGPT

Chat GPT stands for Generative Transform.
Chat GPT is analogous to the results produced on customer service websites - ask questions and get answers.
ChatGPT is an AI program called a chatbot that employs natural language processing to create a humanlike conversational dialogue.

The language model responds to questions and can create various forms of written context, such as articles, messages, emails, computer programs, and media posts. Maybe even artwork.

The natural language processing includes machine translation, information retrieval, sentence analysis, and information extraction.

ChatGPT is a kind of generative AI that involves entering prompts and receiving humanlike images, test, or video in return.

ChatGPT is similar it automated results and services that are found on customer service websites: ask questions and receive clarification.

Chat GPT is trained with reinforcement learning through human feedback and reward systems that analyze and rank the responses. This is machine learning.

ChatGPT is a product of a company named OpenAI.

ChatGPT uses algorithms to find patterns within data sequences.

ChatGPT uses deep learning to produce text with the use of neural networks.

ChatGPT produces text, next word, sentences, or paragraphs based on some learning formulation.

Machine Learning and Types of AI

<u>Artificial Intelligence Specifics</u>

It is the study and development of techniques for making machines that people would class as being smart, especially computer systems and programs. It involves the study of human intelligence, but it is not limited to what we would class as human activity. Thus, the subject of Artificial Intelligence defines the concept of intelligence for its own purpose. Like saying a checkers program is intelligent.

At this point in its development, there are no systems that are more intelligent than humans. There are programs that do very complicated things, however, and they are under the control of other computers or human beings.

The goals of Artificial Intelligence are the development of machines - usually computers - that can think and act like human beings and incorporate thinking and acting accordingly.

Theoretically, the scope of Artificial Intelligence ranges from simple programs to systems that are more intelligent than humans. Many people believe that Artificial Intelligent machines will take over the civilized world, and just as many, perhaps more, think that is not true.

Machine Learning

There are two types of Artificial Intelligence: weak AI and strong AI. *Weak AI is* sometimes referred to as *Narrow AI,* since people like fancy names and also *Artificial Narrow Intelligence* (ANI), that are designed to perform specific small tasks. Some ANI systems are exceedingly simple like vacuum cleaners that can roam around the living room, and a little more complicated self-driving cars and IBM's Watson.

Strong AI is a group of AI's, including Artificial General Intelligence (AGI) or Artificial Super Intelligent (ASI) with intelligence equal to humans. It could do anything, such as solve problems and plan for the future. It would have super intelligence. Strong idea is a good idea, to some degree, but doesn't exist today.

This opens the door to deep learning and machine learning. Since the two names are used interchangeably, it is prudent to denote the difference between them. Machine learning is a sub area of Artificial Intelligence, and deep learning is a sub area of machine learning.

The types of learning differ in how the algorithm learns. Non-deep learning uses human experts to analyze data while deep learning does it by itself.

Deep learning employs neural networks and is generally considered mor of what Artificial Intelligence should be. Neural networks is with Quantum Computing as two of the most complicated subject in advance computer technology.

Generative models take raw data and generate useful information. Generative models take raw data and generate useful information. Generate models are not new and are what the subject of statistics is designed to do. The modern Artificial Intelligence method use information to select data for display or other forms of computation.

Artificial Intelligence Applications

Most Artificial Intelligence writers summarize into 5 categories. They are: Speech Recognition, Customer Service, Computer Vision, Recommendation Engines, and Automated Stock Trading. Here is a brief summary of those categories.

Speech: Uses natural language processing to analyze human speech and generate it in a written form.
Customer: Called virtual agents that respond to customer queries, that may be questions, products, and sizes through customer engagement. Referred to as virtual agents or voice assistants.
Vision: Used to derive useful information from digital images and videos. Used in healthcare, law enforcement, or design by computer.
Recommendation: Use past consumption to discover data trends to improve various business goals.
Automated: Used to optimize stock portfolios and facilitate automated stock trading.

Cognitive Computing

Conventional computing and cognitive computing
 Convention computing is, as practically everyone knows, is a pre-planned set of steps to provide a per-planned result. With convention, applications are conceived, designed, programmed or coded, tested, and then put into operation through a person-oriented scenario. The only intelligence is in the mind of the person's involved. The objectives are identified beforehand, and additionally, computing elements are designed to interact of a pre-defined manner.

 With cognitive computing, elements are defined and combined to execute in an intelligent manner without human intervention. The computing elements may be defined and prepared or exist in the outside operational environment.

Machine Conversation
 Machine conversation is a defined collection of natural language processing and machine learning elements. In its most simple form it is two conversational programs that talk to one another, such as ELIZA and ELIZA or DOCTOR and DOCTOR, mentioned previously in this report. Intelligent machine conversation involves two compatible intelligent entities interactions with one another.

Computer Speaking
 The ability to initiate computer speech is a companion to speech recognition. It is usually combined with speech recognition and cognitive computing. A simple example is text reading and speaking. When combined with other cognitive elements, it is in the AI domain.

Empathy from Computers

There are two kinds of computer empathy: human empathy for computer technology and its various difficulties, and the empathy of computers to humans. Computer can sense emotion if defined beforehand. As yet, it has not been considered in the area of Artificial Intelligence.

Cognitive Systems

The discovery and understanding of unstructured information and cognitive systems is generally combined under the title Generative Models.

A generative model is a deep learning facility that can take raw data, such as database information, speech, visual information, and other event data and draw relevant conclusions.

**END OF CHAPTER SEVENTEEN
AND PART TWO**

Part III

ChatGPT, IBM Watson and Data Science

A Gentle Introduction
(Chapters 18-20)

Chapter 18

What is ChatGPT

ChatGPT stands for Generative Pre-trained Transformer. It's a complex computer system that generates text by responding to a user's input. It is a sophisticated and a well thought computer program in a software category known as Generative Artificial Intelligence (GAI). That's what some people think and others question the intelligent part and refer to it as generative software. It has received nationwide attention and more than 200 books have already been written on the subject. Some people love it and others think it will take over civilization, as people refer to it, and we will be extinct intellectually. By the way, that is called *Singularity*. You can go online to its developer, *OpenAI*, to use it. It's free and easy to use. The concept is popular, and you can even try an app for your iPhone that will also generate generative output. This information is given in this part of the book.

Introduction

ChatGPT is an Artificial Intelligence Chatbot that uses natural language processing to create a human like conversational dialogue. The language model can respond to questions and compose various written content, such as articles, social media posts, essays, computer programs, and messages and emails.

ChatGPT employs natural language processing concepts, such as machine translation, information retrieval, sentiment analysis,

and information extraction. Accordingly, ChatGPT is a form of generative AI wherein the user enters prompts and receives humanlike images, text, or video that are created with AI methods.

ChatGPT generates results that are similar to the automated output of customer service websites; the user asks questions or receives clarification of some sort and receives a response. With ChatGPT, and not necessarily with sister systems, the system uses the previous results to progress to the next level. As such, the ChatGPT system can create a scenario, written text, such as a novel, or conventional written material of practically any nature. Purportedly, ChatGPT can work with text, pictures, video, or any form of input.

During operation, ChatGPT uses internal scenarios to assist in generating appropriate results. The results may or may not be relevant to the intent of the user. Most users are enamored with ChatGPT results and there has been a distinct increase of persons involved. For example, there are already more the 200 written and published books on the subject.

Using ChatGPT

ChatGPT uses algorithms to find patterns within data sequences by using language models named GPT-3 and GPT-4 and ordinary GPT. Information is not available on them. GPT stand for Generalized Pre-trained Transformer. You can run ChatGPT by entering:

https://chat.openai.com/blog/chatgpt

and you obtain a user ID and password. You are directed to facilities for using the ChatGPT program. Returning users need only to enter:

https://chat.openai.com

The information you enter is called a *prompt*. Follow the prompts and you are in. You are then directed to enter text and receive output from ChatGPT. A trial user is allows only 5 prompts per session and 50 per day. You can open a ChatGPT account for a $20 monthly fee and obtain a Premium ChatGPT subscription with access to other OpenAI models.

When you enter a prompt, ChatGPT yields a response computed with a complex neural network. The transformer generates a response and remembers the interaction that is used for the next prompt. Often the results are pure nonsense. Outputs are not planned in any particular manner and often they not worth very much except for amusement. In a business environment, the outputs yielded by the transformer would make sense. There are probably organizations that are currently working on the use of ChatGPT to enhance their operational environment.

ChatGPT Internals

ChatGPT stores pattern of data entered my users and information systems and uses those patterns to develop its output. It uses neural network and probability to do its work. Each adaption of ChatGPT works in its own particular way based on its input. Each system uses a selected GT, that is Generative Transformer, to function and it used the database to which it is attached. As we are going to state in the conclusion, you can't just hire a summer student and expect an operational environment. The process will probably be similar to the original adoption of computers that used as good set of operational systems and people who knew how to use it.

ChatGPT is not a search engine that takes input requests and thorough various mix and match operations, generates it output. ChatGPT takes input known and generates a response based on its stored patterns and the prompt. The work is performed by a selected transformer, such as two of them named GPT-3 and GPT-4, with varying amount of capacity and computational power. The transformers use a concept known as the *large language model* (LLM) to produce a natural language response to the prompt.

The key element in ChatGPT is a transformer that produces the responses through the use of a neural network of processing nodes similar in thought to the human brain. Neural networks are exceedingly complicated and perform the work of the transformer. Transformers have to be trained and different based on the data, such as test or video. There are differing methods of initiating a chat and responding to its output.

Social Aspects of ChatGPT

Whenever something new pops up, a groups of outsiders also pops up. Some of the areas are regulation, ethics of application, job loss, insurance, and so forth. There will probably some job enhancement opportunities and some decisions to be made by upper management. There will be no singularity.

The Author's Opinion

As with other areas of computer and information systems, some organization, such as IBM with computers and IBM, Oracle, Amazon, and Microsoft in database technology, some company has to step and take the lead. They have to come in and educate employees and install systems. It can't survive in generative artificial intelligence with shotgun advertising and hope for the

123 | A MANAGER'S GUIDE TO ARTIFICIAL INTELLIGENCE CONCEPTS

best marketing. Direct marketing wherein sales people have a jurisdiction and go in and educate the potential customers is necessary. The fact that over two hundred books exist on ChatGPT doesn't mean much since many are independently published. A lot of people are interested in ChatGPT like they were with computer games. ChatGPT is an excellent research area for universities and OpenAI must be a magnificent company. But, ChatGPT has a way to go. By the way, the first and best book on ChatGPT is Pam Baker's *ChatGPT for Dummies,* published by John Wiley and Sons, Hoboken, NJ, 2023. She assumes that you are 'smart and pressed for time and therefore want all meat and no fluff in a fast and easy read.'

Next, we will take a look at IBM Watson.

END OF CHAPTER EIGHTEEN

Chapter 19

What is IBM Watson

IBM Watson, usually referred to simply as Watson is a question answering system that can answer questions entered in common ordinary natural language. The Watson system first entered the world of technology in a competition on the Jeopardy! TV show against former champions. Watson won. I has since been evolved by IBM into a generative system for supporting end-user computing. Watson was a creative excellent wonderful imaginative project that only IBM could think up and execute with excellence. Watson is an effort that only a team with superior knowledge and fantastic management could pull off.

Introduction

Watson has since been adapted to a CLOUD service accessible to a myriad packages with generative intent. Now this is the author's projection or opinion, depending how you look at it. Then for each class of computer application, you need access and training packages for each application class, as well as user interface modules. This is what IBM has done with their System/360 and System/370 computers to support through microprogramming several basic hardware models operate and be programmed as one computer design. Then, if you combine this with direct marketing, you have something to think about. IBM may already be doing that, how would I know.

Operational Considerations

In this form, Watson works in the same a manner as ChatGPT, and the end-user can look at it in that way. Of course, they are not the same. The user inters a prompt and the system does something. That something could be an important business function or intended output of a visual or textual nature. During operation, the Watson system passes the prompt through a neural network system and generates an appropriate response. In so doing, it can be used to solve problems in the following areas:

- Coding computer programs.
- Composes music.
- Make wireless drafts of emails.
- Summarize articles.
- Writing social media posts.
- Creating articles.
- Making articles.
- Describing complex topics.

This is only a generic list that would be provided with any practically any generative AI systems.

Being an IBM product, Watson would probably avoid the following problems inherent with generative systems:

- Understanding the complexity of human language.
- Lack up-to-date information.
- Behave like a machine.
- Has poor analytics.
- Has trouble understanding the relevancy of specific data.

Watson would appear to be devoid of the given limitations, as stated.

How in Watson Works

The Most important feature of Watson is that it operates out of the IBM Cloud, which is an operational feature that is relevant to modern computer and data processing systems. All resources, except the user interface, reside and operate in the cloud and that is where all support modules and the Watson hardware and software reside. In order to use Watson, you have create an account that necessitates that you have to log in whenever you want to use your facilities. That is an application assistant that gives you the interface between your application and the Watson facilities.

You have to tell the application the *intents* which are the actions it will do for your application. Next, you have to identify *identities* that are the functions the assistants will do for you. As you can so, you are building an application interface. Next a *dialog* has to be built in order for the end user can communicate with your application that runs on then cloud.

When you are dealing with vision, you have to build a *classifier*. So you can see how using the IBM Watson operates. It is deceptively simple to build an application and it is transparent to the user team.

There is a lot of literature on IBM Watson in advanced book form. It is no secret. The reference for the above information is: High, Rob and Tanmay Balshi, *Cognitive Computing with IBM Watson,* Birmingham: Packt, 2019. Rob High is an IBM Fellow.

Working with IBM

When dealing with a system that is being developed, the technological environment is pretty chaotic. Research people at various universities, scientists, professors, and students, along

with capstone projects, theses, and dissertations in addition to the usual technological academic papers and reports. It's no wonder that there more than 200 books published on ChatGPT, many of which are independently published. *OpenAI, the creator of ChatGPT, has done a miraculous job with their development*

With IBM Watson, you have to buy into IBM, so to speak, and you are assisted in all phase of the project.

END OF CHAPTER NINETEEN

Chapter 20

What is Data Science

Data science is discipline that helps to explore the existence and manipulation of data for business. The author believes that Data Science can be useful for any organization, the government, and individuals. It employs the fields of mathematics, statistics, artificial intelligence, and computer science. The worldwide upsurge of interest in Artificial Intelligence parallels the interest in Data Science. The objective is explore large amounts of data to investigate what has happened, why it happened, and what could be the future results that are implied by that data. As a discipline, it explores the meaning inherent in data. Human beings employ data science of their own in everyday life, and Data Science does that with large amounts of data.

Introduction

Data science first appeared and a name for applied statistics. It answers the question asked by many students, "Why are we studying this?" Modern organizations are constantly in the process of collecting large amount of information. E-Commerce, medicine, finance, and so on, just to conduct their everyday business. Of course, they require data to perform operations, but isn't there a good use other than operations for this data. The answer is if we investigate the aspects of data design, collection, and analysis, we can get an added benefit to the mountains of data that are getting higher and higher.

AWS, which stands for Amazon Web services, had looked into data analysis and has come up with the following areas of investigation"

- Descriptive analysis
- Diagnostic analysis
- Predictive analysis
- Prescriptive analysis

The objective of *descriptive analysis* is to find out what is going on in the organization. *Diagnostic analysis* is to what out why it happened. *Predictive analysis* uses history to predict what will happen. *Prescriptive analysis* predicts what is going to happen, and employs simulation, neural networks, artificial intelligence, and machine learning.

Benefits

The benefits of Data Science is to uncover hidden patterns hidden in the massive amount of data in many organization. It is good way to initiate new products and optimize the current organization.

Here is how Data Science operates: obtain data, scrub data, explore data, model data, and interpret results. It uses the techniques of classification, regression, and clustering. This is why personnel educated in Data Science are required.

Technology

This is where relevant technology and organization of resources comes inn. Paired with artificial intelligence, you are going to explore cloud computing, such and IBM Watson, and maybe quantum computing in the future.

Data Science is not something for the future, it is something for the present. A reference is Mueller, J and L. Massaron, *Data Science Programming,* Hoboken, NJ: Wiley, 2020.

**END OF CHAPTER TWENTY
PART THREE AND THE BOOK**

ABOUT THE AUTHOR

Harry Katzan, Jr. is a professor who has written several books and many papers on computers and other subjects. He has been an advisor to the executive board of a major bank and a general consultant on other disciplines. He and his wife have lived in Switzerland where he was an AI consultant and a visiting professor. He holds bachelors, masters, and doctorate degrees.

BOOKS BY HARRY KATZAN, JR.

Computers and Information Systems

Advanced Programming
APL Programming and Computer Techniques
APL Users Guide
Computer Organization and the System/370
A PL/I Approach to Programming Languages
Introduction to Programming Languages
Operating Systems
Information Technology
Computer Data Security
Introduction to Computer Science
Computer Systems Organization and Programming
Computer Data Management and Database Technology
Systems Design and Documentation
The IBM 5100 Portable Computer
Fortran 77
The Standard Data Encryption Algorithm
Introduction to Distributed Data Processing
Distributed Information Systems
Invitation to Pascal
Invitation to Forth
Microcomputer Graphics and Programming Techniques
Invitation to Ada
Invitation to Ada and Ada Reference Manual
Invitation to Mapper
Operating Systems (2nd Edition)
Local Area Networks
Invitation to MVS (with D. Tharayil)
Introduction to computers and Data Processing
Privacy, Identity, and Cloud Computing

Business and Management

Multinational Computer Systems
Office Automation
Management Support Systems
A Manager's Guide to Productivity, Quality Circles, and Industrial Robots
Quality Circle Management
Service and Advanced Technology

Basic Research

Managing Uncertainty
Microprogramming Primer

Service Science

A Manager's Guide to Service Science
Foundations of Service Science
Service Science
Introduction to Service
Service Concepts for Management
A Collection of Service Essays
Hospitality and Service

Little Books

The Little Book of Artificial Intelligence
The Little Book of Service Management
The Little Book of Cybersecurity
The Little Book of Cloud Computing
The Little Book of Managing Uncertainty

Novels

The Mysterious Case of the Royal Baby
The Curious Case of the Royal Marriage
The Auspicious Case of the General and the Royal Family
A Case of Espionage
Shelter in Place
The Virus
The Pandemic
Life is Good
The Vaccine
A Tale of Discovery
The Terrorist Plot
An Untimely Situation
The Final Escape
Everything is Good
The Last Adventure
The Romeo Affair
Another Romeo Affair
Understanding

END OF BOOKS BY HARRY KATZAN JR.

.

www.ingramcontent.com/pod-product-compliance
Lightning Source LLC
LaVergne TN
LVHW011946070526
838202LV00054B/4817